AF352864

Impact of Agricultural Practices on Biodiversity of Soil Invertebrates

Impact of Agricultural Practices on Biodiversity of Soil Invertebrates

Editors

Stefano Bocchi
Francesca Orlando

MDPI • Basel • Beijing • Wuhan • Barcelona • Belgrade • Manchester • Tokyo • Cluj • Tianjin

Editors
Stefano Bocchi Francesca Orlando
University of Milan University of Milan
Italy Italy

Editorial Office
MDPI
St. Alban-Anlage 66
4052 Basel, Switzerland

This is a reprint of articles from the Special Issue published online in the open access journal *Agronomy* (ISSN 2073-4395) (available at: https://www.mdpi.com/journal/agronomy/special_issues/Soil_Invertebrates).

For citation purposes, cite each article independently as indicated on the article page online and as indicated below:

LastName, A.A.; LastName, B.B.; LastName, C.C. Article Title. *Journal Name* **Year**, *Volume Number*, Page Range.

ISBN 978-3-03943-719-1 (Hbk)
ISBN 978-3-03943-720-7 (PDF)

Cover image courtesy of Valentina Vaglia.

Contents

About the Editors

Stefano Bocchi is a Full Professor of Agronomy and Cropping Systems at the Department of Environment Science and Policy, State University of Milan, Italy. As a Visiting Scientist at the Agronomy Department University of California, Davis, IRRI – Philippines, and Wageningen University, he developed research projects on cereals, forage crops, agro-food systems analysis, and management. He is a teacher in several graduate and undergraduate courses of agronomy, agroecology, agro-food systems, foodshed management, landscape agronomy, tropical crops, and organic farming; member of the scientific board of the Ph.D. program in agro-ecology; director of the "Geomatic Lab for Agriculture and Environment" for agricultural and cropping system analysis, biodiversity, forest and pasture inventories, use and application of GIS tools for natural and agricultural resource analysis; Director of CICSAA Inter-University Centre for International Cooperation for Agro-food systems development; team research leader in several projects on farming system analysis and management, both at national and international level. He has been involved in several projects for international cooperation in Albania, Brasil, China, Ecuador, Egypt, Kenya, Lebanon, Peru, Philippines, Sierra Leon, and Tanzania. He is the delegate of the Dean for sustainability. He is the author of more than 170 scientific papers and a Board Member of various scientific societies. He was the scientific curator of the Biodiversity Park in EXPO 2015.

Francesca Orlando was a researcher at the Department of Environmental Science and Policy (University of Milan) and then at the Department of Molecular and Translational Medicine (University of Brescia). During her work, she addressed different aspects of the integrated assessment of sustainability in agriculture. She obtained her degree in Agricultural Science, and a Ph.D. in Climatology and Soil Science, from the University of Florence. She is an expert in agronomy and soil science, with a focus on the impacts and benefits of the agricultural practices belonging to conservative agriculture, precision farming, and organic farming. During her career, she studied the implications of different crop management methods and climate and micro-climate trends in chemical, physical, and biological soil fertility, as well as on the quality and quantity of the productions. She was involved in many national and international projects and was WP leader of 3 projects concerning precision farming and organic farming for rice. She has given numerous presentations at international conferences and published about 30 papers focused on the performance and efficacy of innovative techniques and tools in agriculture, aimed to optimize input use and reduce the environmental impacts of agriculture on natural resources, such as soil.

Preface to "Impact of Agricultural Practices on Biodiversity of Soil Invertebrates"

Soil fauna plays a key role in many soil functions, such as organic matter decomposition, humus formation, and nutrient release, modifying the soil structure and improving its fertility. In particular, soil invertebrates play key roles in determining soil suitability for agricultural production and realizing sustainable farming systems. This fauna includes an enormous diversity of arthropods, nematodes, and earthworms. However, they suffer from the impact of agricultural activities, with implications for the capacity of soil to maintain its fertility in the long-term and to provide ecosystem services. Some agricultural practices may create crucial changes in soil habitats and properties, with consequences for invertebrate biodiversity. In the few last decades, especially under intensive and specialized farming systems, a loss in soil ecosystem services has been observed, as a result of the reduction in the biological fertility with a decrease of both the abundance and taxonomic diversity of the soil faunal communities.

On the other hand, some agricultural practices, such as those based on crop rotation, minimum tillage, cover crops and soil-covering, can support sustainable soil management and promote useful soil fauna. Therefore, due to concerns about the sensibility of soil biota to the agricultural practices, there is an urgent need to develop sustainable management strategies, to realize a microclimate and habitats favorable for the biodiversity of soil invertebrates and to reduce the soil disturbance, such as that due to tillage or chemical input.

Stefano Bocchi, Francesca Orlando
Editors

Article

Soil Arthropod Responses in Agroecosystem: Implications of Different Management and Cropping Systems

Cristina Menta [1,*], Federica Delia Conti [1], Carlos Lozano Fondón [1], Francesca Staffilani [2] and Sara Remelli [1]

1   Department of Chemistry, Life Sciences and Environmental Sustainability, University of Parma, Viale delle Scienze 11/A, 43124 Parma, Italy; conti.federica@gmail.com (F.D.C.); lzncls@unife.it (C.L.F.); sara.remelli@unipr.it (S.R.)
2   Geological, Seismic and Soil Survey-Emilia-Romagna Region, Viale della Fiera, 8, 40127 Bologna, Italy; Francesca.Staffilani@regione.emilia-romagna.it
*   Correspondence: cristina.menta@unipr.it

Received: 9 May 2020; Accepted: 6 July 2020; Published: 9 July 2020

Abstract: The EU's Common Agricultural Policy (CAP 2014–2020) on soil management points to the combination of sustainable food production with environmental protection, reduction of CO_2 emissions, and safeguarding of soil biodiversity. In this study, three farms (in the Emilia-Romagna region), managed with both conventional and conservation practices (the last ones with and without sub-irrigation systems), were monitored from 2014 to 2017 to highlight the impact of different crops and soil managements on soil arthropods, in terms of abundance, composition, and soil biological quality (applying QBS-ar index). To do this, linear mixed models were performed, whereas arthropods assemblages were studied through PERMANOVA and SIMPER analysis. Soil communities varied among farms, although most differences were found among crops depending on management practices. Nonetheless, conservation systems and a wider reduction in anthropogenic practices provided better conditions for soil fauna, enhancing QBS-ar. Moreover, arthropod groups responded to soil practices differently, highlighting their sensitivity to agricultural management. Community assemblages in corn and wheat differed between managements, mainly due to Acari and Collembola, respectively. In conservation management, wheat showed the overall greatest abundance of arthropods, owing to the great number of Acari, Collembola, and Hymenoptera, while the number of arthropod groups were generally higher in crop residues of forage.

Keywords: soil biodiversity; bioindicators; soil quality; mesofauna; soil degradation; land management

---

1. Introduction

Soil provides the basics for human livelihood and well-being, including food supply, freshwater and many others ecosystem services, in addition to biodiversity [1]. This is especially the case with the soils of agricultural areas, which account for 13% of the total ice-free land cover at the global scale, and are amongst the most important resources for ecosystem functioning, often compromised by mismanagement. Biodiversity plays a crucial role in ecosystem functioning and services [2]; nevertheless, many authors have highlighted the negative effects of conventional management on soil biodiversity multifunctionality [3]. Practices such as tillage, overfertilization, monoculture, and pesticide application often give rise to increased soil erosion, decay of organic matter content, salinization, and compaction, which may lead to a reduction in crop productivity and soil biodiversity, and subsequent socioeconomic losses [3,4]. In the past decade, research on conservation practices such as long-term tillage, diversification of crop production systems, rotation, and crop perennialization has proven to enhance the stabilization of soil organic matter aggregates [5], enhancement of crop yields, improvement of carbon sequestration, nutrient retention, and water infiltration [4].

Soil fauna plays an important role in maintaining soil quality and health, as well as providing ecosystem services [6] through processes such as organic matter translocation, fragmentation and decomposition, nutrient cycling, soil structure formation and, consequently, water regulation [7–9]. Some groups are highly sensitive to changes in soil quality because they are adapted to specific soil conditions [10,11]. Among soil fauna, mesofauna (200 μm–2 mm) are affected by both above- and belowground environmental factors since their activity occurs mostly in the top 20 cm of the soil profile. Within aboveground factors, mesofauna presence is affected by plant cover, which sensibly impacts soil properties by shading and regulating soil temperature, allowing steady environmental conditions. Moreover, plants generate litter inputs, thereby enhancing soil hydrophobicity and protecting it from erosion. Within belowground factors, instead, rhizosphere organic compounds that involve the root exudates represent food resources for soil mesofauna. However, due to their small dimensions, these organisms use existing pores or channels for locomotion, which makes many of them sensitive to any interference with the soil environment. Within the habitable pore space, their activity is influenced by water-air proportion, such that both saturation and desiccation processes, resulting in anaerobiosis and dehydration respectively, are detrimental to soil fauna populations [12]. Cole et al. [13] concluded that communities inhabiting agroecosystems are primarily structured by agricultural practices, since anthropogenic activities of agricultural systems alter natural soil dynamics and promote the decay of soil mesofauna populations.

Soil management practices lead to alteration of plant litter inputs and soil microhabitat, in terms of both soil physical and chemical qualities, thus impacting soil fauna assemblages [14]. It has been widely observed that tillage impacts negatively on soil-dwelling arthropod communities: enhancing the exposure of soil organisms to desiccation, through the destruction of upper horizons; and negatively affecting access to food sources, through the decrease in the soil moisture available and the disruption of existing plant systems [15–17]. Conversely, no-tillage practices leave the soil surface covered with residues of previous crops, thereby protecting the soil from water and wind erosion and enhancing decomposer fauna abundance [18]. These effects increase in continuous no-tillage systems, due to the higher soil stratification and concentration of organic matter, nutrients, and microbial activity near the surface [19]. Several studies found that even mite, generally widespread in soils under no-till practices or uncultivated soils, are negatively affected by conventional tillage, especially those belonging to Oribatida [20–22]. Cortet et al. [15] observed a reduction by more than 50% in the number of Acari in tilled soil. On the other hand, even if it has been widely accepted that Collembola are highly discriminant among agricultural management systems, the response is inconsistent between studies. For example, Filser et al. [23] found a higher abundance of Collembola associated with intensive and moderate systems compared to sustainable systems, suggesting that springtails can create large populations under high management intensity; while Maraun et al. [24] suggested that Collembola are sensitive to mechanical disturbances, even more than oribatid mites. Not only Collembola, but also Isopoda and Pauropoda meet a significant reduction caused by mechanical and chemical perturbations produced in conventional agricultural management practices [25,26]; for instance, Palacios-Vargas [27] noted that Pauropoda are very sensitive to agricultural practices, the impact of which is demonstrated to have reduced their populations by about 70%. Moreover, since Symphyla are negatively affected by high bulk density, soil compaction through tillage practices is suggested to influence their occurrence as well [26].

Another decisive factor, when assessing the effect of management practices on soil arthropods, is crop rotation [18]. Meyer et al. [28] discovered an ecological 'memory effect' in the soil community, i.e., an influence from the preceding crops on soil fauna composition, suggesting that crop rotation can be a useful tool to increase arthropod biodiversity and biomass, particularly in areas managed with monocultures. Actually, Jones et al. [29] observed that monoculture cropping system is considered a cause of decrease in the diversity of microhabitats, thus of Isoptera.

On the other hand, organic fertilizers, like manure, are generally beneficial to almost all soil organisms, even though these beneficial effects may be partly due to the high number of soil arthropods found in the manure itself [30]. Nevertheless, the dosage could affect the outcome of the fertilizer and occasionally produce negative effects [31]. On the other hand, Kautz et al. [32] suggested that the abundance of soil arthropods increases differently depending on regimes of organic manuring; for example, they found that the application of straw and green manure increased the abundance of soil arthropods, contrary to mineral nitrogen. They also highlighted that fertilizer application could not take enough time for a significant induced modification of the fauna composition, so only effects on abundance may be observed. Nevertheless, Cluzeau et al. [33] noted that Collembola abundance increased with both mineral and organic fertilization, while mites tend to take more advantages by fertilisation with manure than using mineral fertilisers alone [34].

Changes in species diversity can be observed between crops, depending on plant physiology and consequently on biochemical processes and metabolism. These properties acquire much more importance since the carbon intake of below-ground soil fauna could derive from roots, other than from litter [35]. For example, compared to sugar beets, wheat plants generally possess a more complex system of roots and associated microorganisms, along with greater quantities of exudates [36]. Thereby, wheat rhizosphere could provide a more diversified food base for soil organisms. Sticht [37] suggested that these different nutritional conditions affect the community composition and dominance distribution of collembolans, leading to greater diversity under winter wheat. This result is supported by other organisms such as carabid fauna: Holland and Luff [38] found that although no species has been linked with a particular crop plant, the greatest difference occurs between winter sown crops (namely, cereals and oilseed rape) and spring sown root crops (namely, potatoes, sugar beet, maize, carrots), with the latter ones usually having lower abundance and diversity. Moreover, this study also suggests that differences at the species level could be a consequence of the microclimate established in root and cereal crop systems, it being much drier and warmer in cereals.

In order to protect natural resources and environment, the Emilia-Romagna Region (located in north-central Italy, between the Po River and the Apennine Mountains) has adopted soil conservation management practices. The aim is to achieve production with less pesticides, chemicals, and water inputs and reductions in CO_2 emissions, as required by the EU's Common Agricultural Policy (CAP 2014–2020). To support this policy, in 2013, Emilia-Romagna and four other northern Italian regions (Friuli Venezia Giulia, Veneto, Lombardy, and Piedmont), set up a Life project called HelpSoil, which ended in June 2017. This project was financed by the European Commission and aimed to compare soil management practices of conservation agriculture (no/minimum tillage, permanent soil cover) with conventional plowing-based techniques on twenty demonstration farms located in the five regions. In addition to the Life HelpSoil project goals, additional soil samples were collected on three chosen farms of the Emilia-Romagna, in order to improve understanding of soil arthropod communities in different agricultural systems. This paper reports the results of this study, which aimed to evaluate how different crops and soil management practices affect soil arthropod communities, in terms of both abundance and composition, as well as soil biological quality using QBS-ar, index applied on soil arthropod community [39]. In detail, our hypotheses are: (i) comparing the effects of conventional (higher disturbance) and conservation (lower disturbance) farming practices on soil arthropod communities, we assume that minimum disturbance increases soil arthropods abundance and diversity; (ii) highlighting the effects of cover crops on soil arthropod communities between and within agronomic and sub-irrigation systems, we assume that permanent soil cover can promote soil arthropods. At the end, individualization of the most sensitive arthropod taxa (at orders or class level) to agricultural practices, may target potential indicators of soil health in agricultural ecosystems.

2. Materials and Methods

2.1. Farm Characteristics and Soil Management

In the present study, the fields belonging to three farms (Ruozzi, Gli Ulivi, and Cavallini) located in the Emilia-Romagna region (Northern Italy) were monitored from 2014 to 2017 through four sampling periods (autumn 2014, spring and autumn 2015, spring 2017). The three farms are inserted in an intensive agriculture scenario characterized by cereal, cereal-forage crop and fruit production, where soil threats and environmental issues are different due to different pedo-climate conditions (Figure 1).

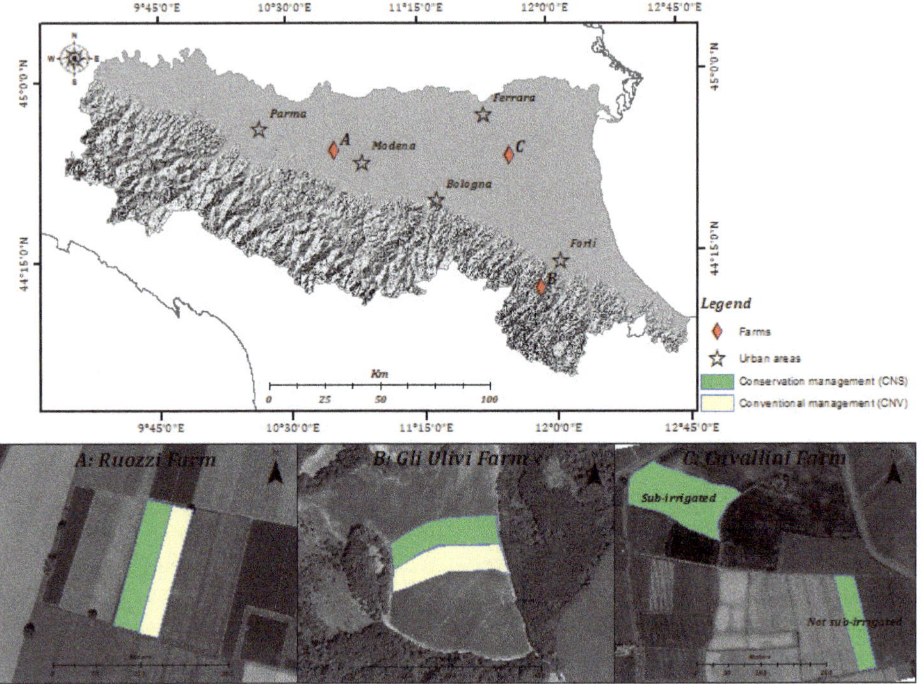

Figure 1. Geographical position of the three farms in the Emilia-Romagna region, using ArcGIS (version 10.4.1) and Google Earth Pro (v. 7.3.2.5776): (**A**): Ruozzi farm; (**B**): Gli Ulivi farm; (**C**): Cavallini farm [40–43].

Two management types, i.e., conventional (CNV) and conservation (CNS), were compared on both Ruozzi and Gli Ulivi farms. Conservation practices only were adopted on Cavallini farm. In this farm, the presence/absence of sub-irrigation was compared. Soil data were determined according to the official methods of soil analysis in Italian legislation (DM 13/09/1999 SO n.185; Table S1) [44]. Crop types for each sampling period in the three farms—Ruozzi, Gli Ulivi, and Cavallini—are shown in Table 1.

The Ruozzi agro-zootechnical farm is located in the Po alluvial plain, Reggio Emilia province (Figure 1A). The soil is classified as fine, mixed, active, mesic Vertic Calciustept (USDA 2010), and is characterized by a high percentage of clay up to 100 cm deep; consequently, it is subject to cracking in the dry period and is very adhesive and plastic when wet (Table S1). One of the main problems of this farm is to ensure good soil drainage and the use of innovative techniques for manure fertilization, aimed at reducing ammonia emissions. The land was used for forage crops (feed wheat from October 2013 to May 2014), followed by a short annual forage crop sown after wheat and harvested in August

2014, and cereal (corn from April to September 2015 and wheat from October 2015 and June 2016); no-tillage practice was adopted under the conservation management, and the permanent cover of soil was ensured through cover-crops (from September 2014 to April 2015) and crop residues during winter 2016/2017. At the sampling periods, the fields (both conventionally and conservatively managed) were covered with crop residues after the annual forage crop in September 2014; corn in June 2015; wheat in November 2015; bare soil after tillage in the conventional system, and crop residues of wheat in the conservation system in March 2017 (Scheme S1A).

Table 1. Crop types for each sampling period in the three farms: Ruozzi, Gli Ulivi, and Cavallini.

Sampling Period	Management	Irrigation System	Ruozzi	Gli Ulivi	Cavallini
2014-Autumn	CNV	NS	After annual forage crop, soil covered by crop residues	Alfalfa	-
	CNS	NS	After annual forage crop, soil covered by crop residues	Alfalfa	Cover crop
		S	-	-	Dried weed
2015-Spring	CNV	NS	Corn	Wheat	-
	CNS	NS	Corn	Wheat	Soybean
		S	-	-	Soybean
2015-Autumn	CNV	NS	Wheat	Bare soil	-
	CNS	NS	Wheat	Cover crop	Cover crop seeding
		S	-	-	Wheat seeding
2017-Spring	CNV	NS	Bare soil	Wheat	-
	CNS	NS	After wheat, soil covered by crop residues	Wheat	After soybean, soil covered by crop residues
		S	-	-	After wheat, soil covered by crop residues

CNV: Conventional management, CNS: Conservation management, NS: No Sub-irrigation system, S: Sub-irrigation system.

The Gli Ulivi agro-zootechnical farm is located on the hills of the Romagna Apennines (in the south-east of the E-R region) (Figure 1B). The soil is classified as fine loamy, mixed, superactive, mesic Typic Haplustept (USDA 2010), and is characterized by medium-texture calcium carbonate aggregations, low content of organic carbon, and a 10 to 40% slope (Table S1). Due to the slope, the soil is difficult to work and is vulnerable to water erosion; consequently, fertility loss is one of the main problems of this place. The land was used for forage crops (alfalfa at its 4th year of vegetation until October 2014) and cereal (wheat in 2014/2015 and 2016/2017, sorghum from May to October 2016); no-tillage practice was adopted under the conservation management, and the permanent cover of soil was ensured through cover-crop (from August 2015 to March 2016). At the sampling periods, the fields (both conventionally and conservatively managed) were covered with alfalfa in September 2014; wheat in May 2015; bare soil after tillage in the conventional system and cover crop in the conservation system in November 2015; wheat in March 2017 (Scheme S1B).

The Cavallini farm is a fruit and cereal farm located in the plain of the Po's ancient delta (Figure 1C). The soil is classified as fine silty, mixed, superactive, mesic Typic Calciustept (USDA 2010), and is characterized by medium to moderately coarse texture, low content of organic carbon and rich in carbonate (Table S1). The soil is susceptible to crusting due to the high percentage of silt in the topsoil, while the presence of sand in depth increases the risk of water deficit. On this farm, both test fields considered in this study were subject to conservation management, but only one had a sub-irrigation plant to limit water consumption. The land in the sub-irrigated field (S) was used for wheat (in both 2013/2014 and 2015/2016) and soybean (from May to October 2015); the other field (NS) was for wheat (in 2013/2014) and soybean (from June to October in both 2015 and 2016). No-tillage practice was adopted in either systems, and the permanent cover of soil was ensured through the use of cover-crop (before soybean in both fields) and crop residues (from autumn 2016 to spring 2017). At the sampling periods, the fields were covered with dried weed in the sub-irrigation system, and cover crop in the no sub-irrigation system in October 2014; soybean in both systems in June 2015; wheat seeding and cover crop seeding, in the sub-irrigation system and no sub-irrigation system, respectively, in November 2015; crop residues of wheat and soybean, in the sub-irrigation system and no sub-irrigation system, respectively, in March 2017 (Scheme S1C).

2.2. Soil Sampling and Arthropod Extraction

Each field was sampled four times during the monitoring period (autumn 2014, spring and autumn 2015, spring 2017). Each time, three replicates of soil cores (10 × 10 × 10 cm) were collected from each field, starting of the field center, following a triangle path, and sampling 60 m away from the other two points. Arthropod extraction was performed by Berlese-Tüllgren funnel for 10 days. The extracted specimens were collected and preserved in 75% ethyl alcohol and 25% glycerol by volume. For defining soil biological quality, the QBS-ar index was applied. This index is based on the biological form approach: (1) the arthropod groups classified at class level for Myriapoda, order level for Hexapoda, Chelicerata, and Crustacea; (2) the different adaptation level of specimens belonging to the same taxon; (3) the adults and larvae of Holometabolous insects (Diptera, Coleoptera, Lepidoptera, Hymenoptera) are considered separately, taking into account the different role and soil adaptation of these two stages. Following the QBS-ar protocol, an ecomorphological score (EMI) was assigned to each taxon found, ranging between 1 and 20 depending on their adaptation to soil (1: low adaptation; 20: maximum adaptation) [39,45]. For each replicate, the QBS-ar value was calculated as the result of the sum of the highest EMI values for each taxon [39].

2.3. Statistical Analysis

Given the particular situation of each location (resulting from specific soil properties; Table S1), consequently to the differences in agricultural management, the statistical analysis was carried out separately per farm.

Linear mixed modelling, using lme4 package, was conducted to evaluate the effect of independent factors on the dependent variable [46]. Managements (conservation vs. conventional) and crop types were considered as independent factors for Ruozzi and Gli Ulivi, whilst for Cavallini sub-irrigation (sub-irrigated vs. not sub-irrigated) replaced management in the model. The dependent variables, analysed one at a time, were: total arthropod abundance, number of ecomorphological groups (EMI), QBS-ar value, proportion of the groups with EMI 20, and the abundance of the biological forms highly representing the soil fauna total abundance, so ≥3%, detected on each farm (i.e., Acari and Collembola on the three farms, Hymenoptera on Ruozzi and Gli Ulivi, Symphyla and Diptera larvae on Cavallini). All factors and interactions were modelled as fixed effect, using a within-subject design to account for repeated measures in the fields. The significance of the model compared with others (with different implementation) was evaluated using log-likelihood ratio. Pair-wise comparisons using the least square means were performed with multcompView and lsmeans packages [47] by applying Holm-Sidak correction for multiple interaction comparisons.

Non-metric multidimensional scaling (NMDS), based on Bray-Curtis dissimilarity index, was performed to visualize how patterns (farms and crop type first, and subsequently management and crop type) influenced the grouping of arthropods communities. The results were plotted in an NMDS ordination diagram, fitting them onto the first two axes. Permutational multivariate analysis of variance (PERMANOVA) was used to test for differences in assemblages among the different patterns visualized with NMDS. In each farm, after a significant PERMANOVA test, an analysis of similarity percentages (SIMPER) was used to test which arthropod groups were driving the differences in assemblages within and between managements. Ordination, PERMANOVA, and SIMPER were performed with the vegan package [48].

Square-root and arcsine transformations were applied on count data and proportions, respectively, in order to meet homoscedasticity and normality of the residuals [49]. A p-value ≤ 0.05 was considered significant. All analyses were performed using R (version 3.6.3) [50].

3. Results

Overall soil arthropod abundance ranged between 382 and 44,222 ind./m^2 of soil. The highest density was observed on Ruozzi farm, in the field conservatively managed under wheat (November 2015), whereas the lowest value was observed on the same farm, but in the field conventionally managed with corn (June 2015; Table S2A). Twenty biological forms were extracted in total, with a minimum of three on the Cavallini farm in the sub-irrigated field after wheat seeding (November 2015; Table S2C), and a maximum of thirteen on Gli Ulivi farm in cover crops (November 2015) in the conservatively managed field (Table S2B). The most abundant groups were Acari (44%), Collembola (37%), Hymenoptera (12%), Coleoptera larvae (1%), Diptera larvae (1%), and Symphyla (1%), accounting for approximately 97% of the organisms collected. Other groups such as Hemiptera, Coleoptera adults, Psocoptera, Araneae, Chilopoda, Pauropoda, and Protura comprised >2%; the remaining taxa were Diplopoda, Thysanoptera, Diplura, Isopoda, Diptera, and Lepidoptera larvae, and reached totally <1%. Among the groups, Acari and Collembola were ubiquitous, but their abundances varied greatly depending on soil management and crops. Both crop typology and farm, as well as their interaction, were correlated with community assemblages, as confirmed by PERMANOVA ($p \leq 0.001$, for both factors and their interaction; Figure 2), so the following analyses were performed separately for each farm.

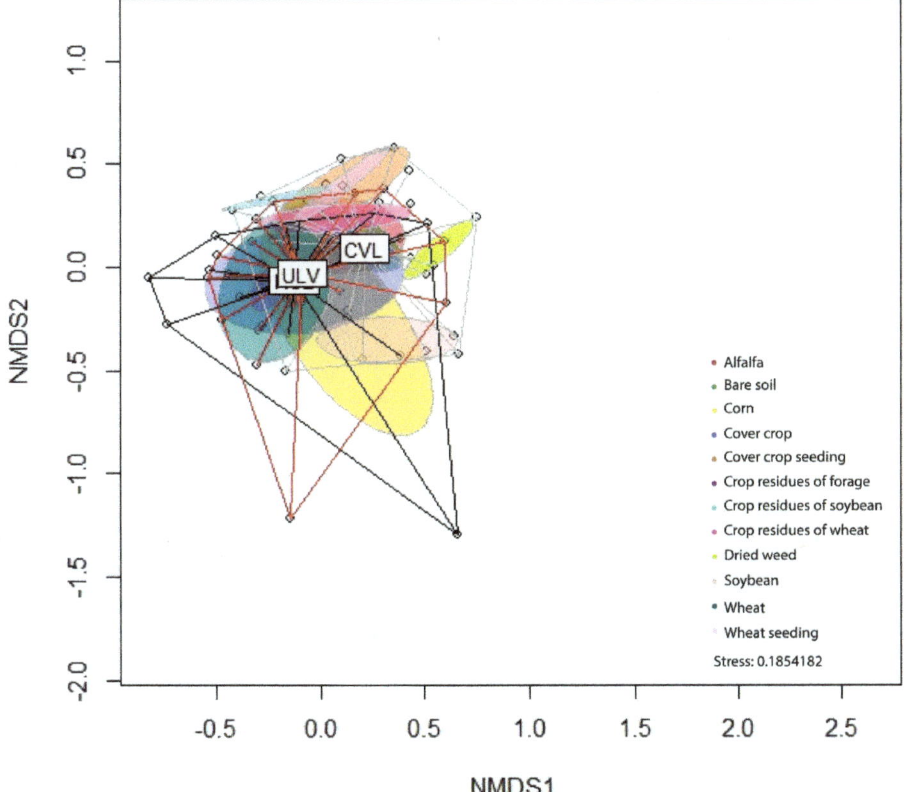

Figure 2. Bray-Curtis based NMDS plot of the arthropod community composition. Points represent samples. "Spider" diagrams connect each point to the belonging farm: ROZ: Ruozzi farm, in black, ULV: Gli Ulivi farm, in red, and CVL: Cavallini farm, in grey. Ellipses represent crop variables.

3.1. Ruozzi Farm

Differences were observed between crop typology as regards the total abundance of soil arthropods ($p < 0.01$; Figure 3A). Contrary, no differences were highlighted between CNV and CNS.

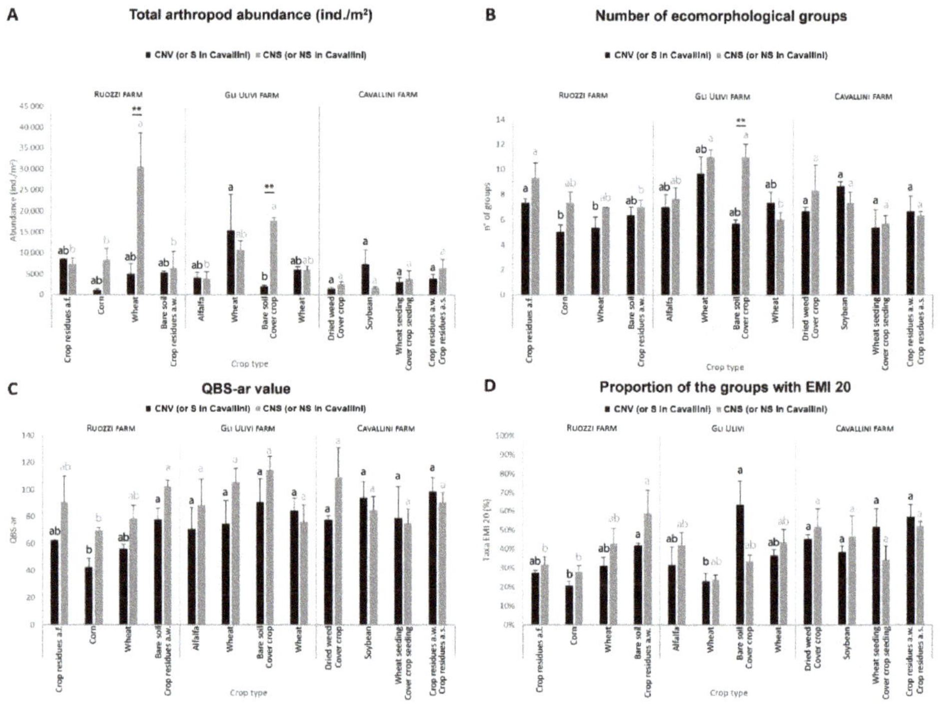

Figure 3. Barplots and standard error of (**A**): total arthropod abundance (ind./m^2); (**B**): number of ecomorphological groups; (**C**): QBS-ar value; (**D**): the proportion of groups with EMI 20. Results for crop types in the three farms are shown; after annual forage crop (a.f.), after wheat (a.w.) and after soybean (a.s.). In the legend, CNS: conservation (or NS: no sub-irrigation system) and CNV: conventional management (or S: sub-irrigation system). Within management, significance ($p \leq 0.05$) is indicated by different letters. Between management, significance is indicated with asterisks: ** $p \leq 0.01$.

However, this variable appeared to be affected by the interaction between management and crops ($p < 0.01$); a difference was indeed highlighted between CNV and CNS in wheat (November 2015) In CNS, soil arthropod abundance in wheat was higher than in the three other crops. In CNV, the different crops showed no significant differences. The number of groups was significantly influenced by the type of crops ($p < 0.01$; Figure 3B): in CNS, it was significantly higher in crop residues after annual forage crop (September 2014) than in crop residues of wheat (March 2017); even in CNV, crop residues after forage crop (September 2014) were higher than corn (June 2015) and wheat (November 2015). The number of groups did not differ between management types. In terms of QBS-ar index, crop type was the only factor affecting the dependent variable significantly ($p < 0.001$; Figure 3C). Within soil management, CNS showed a significant difference between corn (June 2015) and crop residues of wheat (March 2017), higher in the latter, and the same result was highlighted in CNV. The same differences were highlighted for the proportion of EMI-max values (crop type factor only was found significant, $p < 0.001$), where the differences between corn (June 2015) and crop residues of wheat (March 2017) were significant in both managements (Figure 3D). Moreover, the proportion of groups with EMI max found in crop residues of wheat within CNS (March 2017) was significantly higher than crop residues after annual forage crop (September 2014). The community dissimilarity between samples in Ruozzi farm was explained by both management and crop typology, as well as by the interaction between them ($p < 0.001$, for both factors and their interaction; Figure 4).

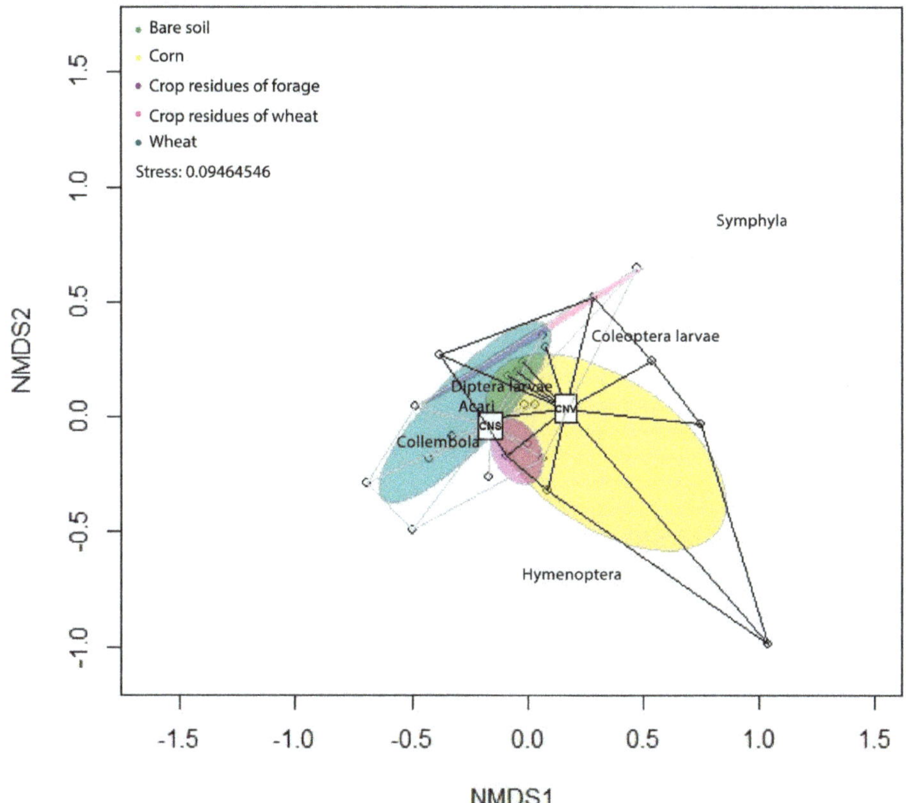

Figure 4. Bray-Curtis based NMDS plot of the arthropod community composition in Ruozzi farm. Points represent samples. "Spider" diagrams connect each point to the belonging management type: CNS: conservation, in grey, and CNV: conventional, in black. Ellipses represent crop variables. To avoid overlapping, only the six more abundant groups were labelled.

Results of SIMPER analysis are shown in Table S3A. Overall dissimilarity within CNV was higher than 50% in all contrasts that involve corn, and was due mainly to Collembola, Acari, Hymenoptera, and Diptera larvae (Table S3(A1)). While, within CNS, an overall dissimilarity higher than 50% was observed only between wheat and crop residues of wheat, with Collembola, Acari, and Hymenoptera accounting for more than 70% of the dissimilarity. Corn and wheat dissimilarities between management were higher than 50% and influenced by Acari, Collembola and, for wheat, Hymenoptera (Table S3(A2)).

On Ruozzi, Acari, Collembola, and Hymenoptera represented the most abundant groups (46%, 45%, and 6%, respectively). Considering the differences within the single group abundance, Acari were observed to be influenced by management, crop type, and the interaction between these two factors ($p < 0.01$, $p < 0.01$ and $p < 0.05$, respectively; Figure 5A).

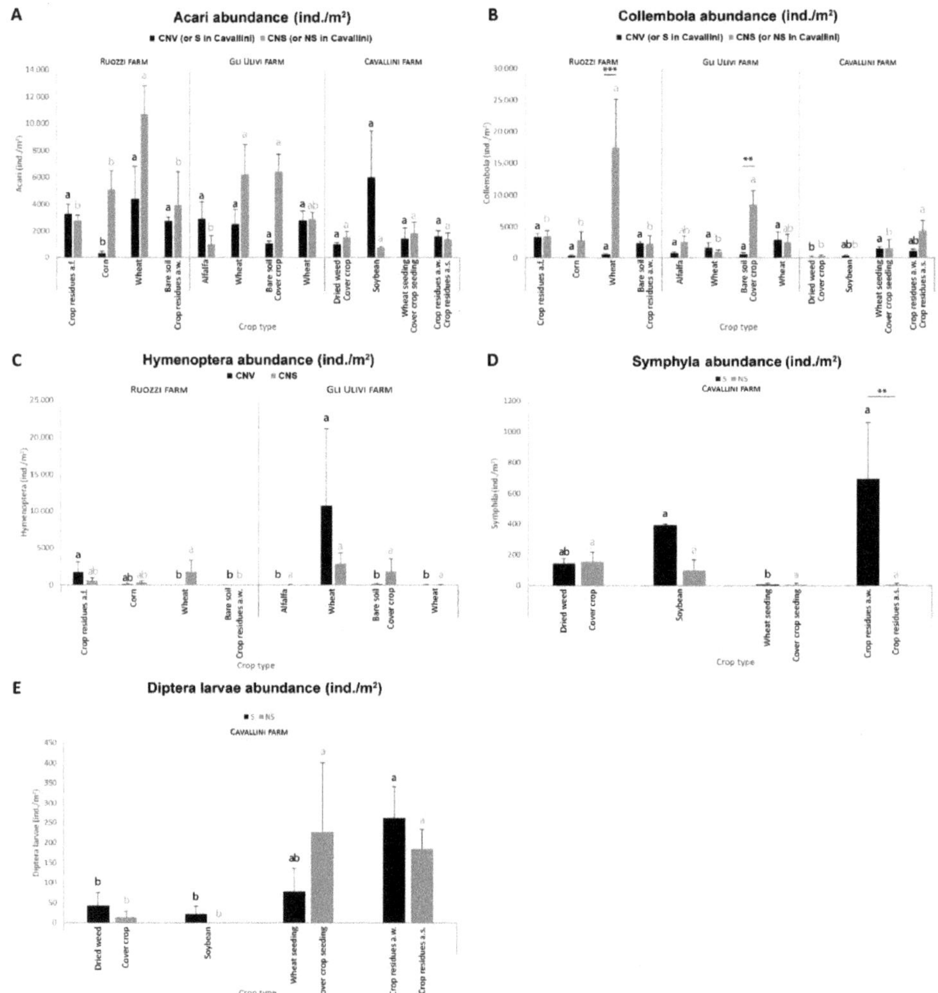

Figure 5. Barplots with standard errors of the abundance (ind./m^2) of the taxa that represent ≥ 3% of the total abundance of the soil fauna found in each farm. In the three farms, (**A**): Acari; (**B**): Collembola. In Ruozzi and Gli Ulivi farms, (**C**): Hymenoptera. In Cavallini farm, (**D**): Symphila; (**E**): Diptera larvae. Results for crop types are shown; after annual forage crop (a.f.), after wheat (a.w.), and after soybean (a.s.). In the legend, CNS: conservation (or NS: no sub-irrigation system) and CNV: conventional management (or S: sub-irrigation system). Within management, significance ($p \leq 0.05$) is indicated by different letters. Between management, significance is indicated with asterisks: ** $p \leq 0.01$.

Although conservation management results in higher Acari abundance than the conventional one, post-hoc comparisons highlighted no difference between a specific combination of management and crop type. However, within the same management, the abundance of Acari was higher in wheat in CNS (November 2015) compared to the other crops. In addition, the Acari abundance was lower in corn in CNV (June 2015) compared to the other crops. Collembolan abundance was affected by both management and crop type, as well as by their interaction ($p < 0.001$, $p < 0.01$ and $p < 0.001$, respectively; Figure 5B). This group highlighted a difference between the two managements in wheat (November 2015). Differences are observed within CNS management, where wheat showed the highest abundance of this group when compared to crop residues after annual forage crop (September 2014, $p \leq 0.01$), corn (June 2015), and crop residues of wheat (March 2017). The Hymenoptera group, on the other hand, was affected by crop type factor only ($p < 0.05$), showing the same trend of Collembola in CNS when comparing wheat with crop residues of wheat (March 2017; Figure 5C). Within CNV management, Hymenoptera displayed the highest abundance in crop residues after annual forage crop (September 2014) when compared to wheat (November 2015) and bare soil (March 2017).

3.2. Gli Ulivi Farm

The total abundance of arthropods in the soils of Gli Ulivi farm highlighted no differences depending on conservation or conventional management only (Figure 3A). On the other hand, there emerged differences depending on crops and their interaction with management type ($p < 0.01$; $p < 0.001$ respectively): a higher abundance in CNS was detected with cover crop (November 2015) when compared to CNV with bare soil (November 2015). Within management type, in CNS under cover crop (November 2015) abundance was higher than under alfalfa (September 2014); while in CNV, higher abundance was observed in wheat (May 2015) when compared to bare soil (November 2015). The number of groups resulted affected both by management and crop type, and by the interaction between the two factors ($p \leq 0.1$; $p < 0.001$; $p \leq 0.001$; Figure 3B). As for the abundance, a higher number of groups was found in CNS cover crop when compared to CNV bare soil (both in November 2015); within CNS, instead, differences were highlighted in the March 2017 wheat, which was lower than the May 2015 wheat, and the November 2015 cover crop. No differences were observed for the QBS-ar comparison, neither between managements systems or different crops within the same management, nor for the interaction of the two factors (Figure 3C). Differently, the proportion of groups with EMI 20 showed differences associated with crops and their interaction with management type ($p < 0.01$ and $p \leq 0.01$, respectively; Figure 3D). Post-hoc analysis highlighted only one difference within conventional management, between wheat (May 2015) and bare soil (November 2015), in which the EMI max was higher in the latter. The significance of management, crop variable, and their interaction were assessed with PERMANOVA after fitting management and crop variables onto community ordination ($p < 0.01$, for management and crop type, and $p < 0.05$, for their interaction; Figure 6).

From SIMPER analysis was observed an assemblage dissimilarity higher than 50% only within CNV, between wheat and bare soil, where Hymenoptera, Acari, Collembola, Psocoptera, and Hemiptera accounted for a cumulative dissimilarity of more than 70% (Table S3(B1)). Between managements, bare soil and cover crop account for an overall dissimilarity of 60%, mostly determined by Collembola, Acari, Hymenoptera, and Diptera larvae (Table S3(B2)).

On Gli Ulivi, Acari, Collembola, and Hymenoptera represented the most abundant groups (39%, 32%, and 23%, respectively). Acari abundance showed differences only in the interaction between management and crop type ($p < 0.01$; Figure 5A). No differences emerged from the post-hoc analysis for the four conditions between the two managements. Within the same management, some differences were detected only in CNS: alfalfa (July 2014) showed lower Acari abundance than wheat (May 2015) and cover crop (November 2015). Like Acari, differences were observed in the interaction between management and crop type for Collembola ($p \leq 0.001$): a higher abundance of collembolans was found in CNS with cover crop compared to bare soil in CNV collected in the same sampling period (November 2015; Figure 5B). Within the same management, only CNS showed a difference—higher in cover crop (November 2015) than in wheat (May 2015). The Hymenoptera abundance appeared to be affected by crop type only ($p < 0.05$; Figure 5C). The abundance of this group was generally low or absent in both managements and in all crops, except for the CNV wheat in May 2015, which was higher than all other crops.

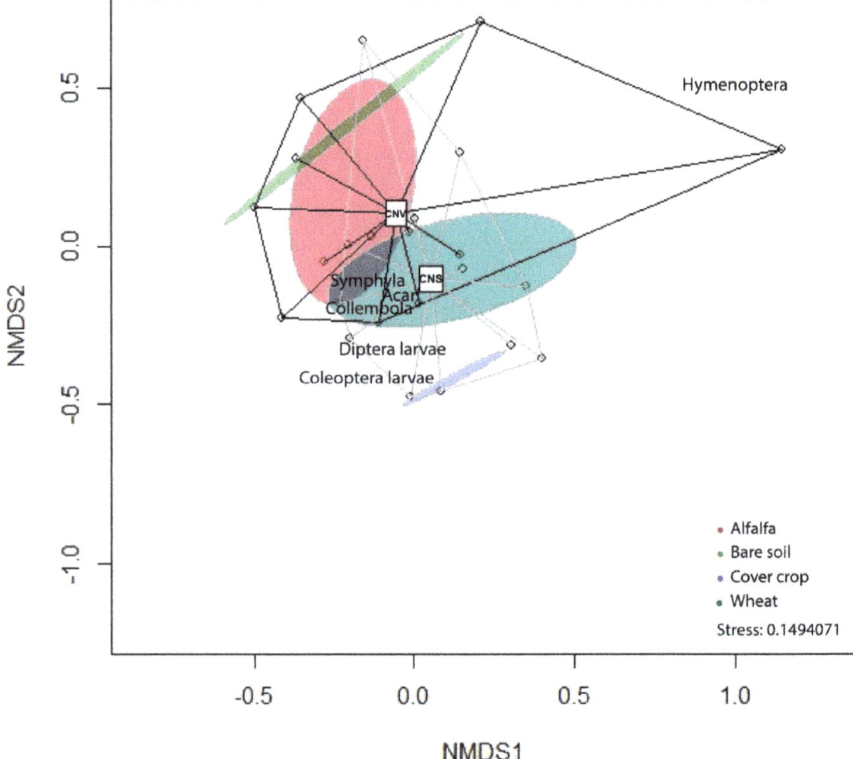

Figure 6. Bray–Curtis based NMDS plot of the arthropod community composition in Gli Ulivi farm. Points represent samples. "Spider" diagrams connect each point to the belonging management type: CNS: conservation, in grey, and CNV: conventional, in black. Ellipses represent crop variables. To avoid overlapping, only the six more abundant groups were labelled.

3.3. Cavallini Farm

On the Cavallini farm, neither total abundance, number of groups and QBS-ar, nor the proportion of groups with EMI 20 appeared to be affected by the presence (S) or absence (NS) of sub-irrigation system and crop type. PERMANOVA revealed an influence of the irrigation system and crop type, and the interaction between them, on arthropods assemblages ($p < 0.001$, for irrigation system and crop type, $p \leq 0.01$, for their interaction; Figure 7).

In Cavallini, within management dissimilarities in arthropods assemblages higher than 50% were observed in S between soybean and wheat seeding, with Acari, Collembola, Symphyla, Hemiptera, and Coleoptera accounting for a cumulative dissimilarity of 71% (Table S3(C1)). Within NS, all contrasts that involve soybean had an overall dissimilarity higher than 50%, as well as the contrast between cover crop and crop residues of soybean, and in both cases, these dissimilarities were due to at least five arthropods groups. Between S and NS, overall differences in arthropods assemblages were less than 50% for all crop types (Table S3(C2)).

Moreover, the community composition on Cavallini farm differed from the ones of Ruozzi and Gli Ulivi (Figure 2). The most abundant groups on Cavallini farm were Acari, Collembola, Symphyla, and Diptera larvae (52%, 31%, 5%, and 3%, respectively). Acari abundance appeared to be affected by the interaction between sub-irrigation system and crop type ($p < 0.05$; Figure 5A). Post-hoc analysis, however, highlighted no specific combination of the two factors. Collembolans showed an influence of crop type ($p < 0.001$; Figure 5B). Within NS, soil with crop residues (March 2017) resulted in higher

Collembola abundance than the other crop types, i.e., cover crops (October 2014), soybean (June 2015), and cover crop seeding (November 2015). Within S, instead, differences were highlighted between wheat seeding (November 2015) and dried weed (October 2014, with higher abundance of collembolans in the former. Symphyla abundance appeared to be affected both by crop type and its interaction with the sub-irrigation system ($p < 0.001$ both; Figure 5D). In March 2017, crops residues in S showed a higher abundance of Symphyla than the NS ones, whilst within S, the abundance in the wheat seeding (November 2015) resulted lower than in soybean (June 2015) and in crop residues (March 2017). Diptera larvae were affected by crop type only ($p < 0.001$; Figure 5E). Where sub-irrigation was absent (NS), cover crop (October 2014) and soybean (June 2015) showed lower results than cover crop seeding (November 2015) and crop residues (March 2017). In the presence of sub-irrigation (S), the number of Diptera larvae was higher in crop residues (March 2017) compared to cover crop (October 2014) and soybean (June 2015).

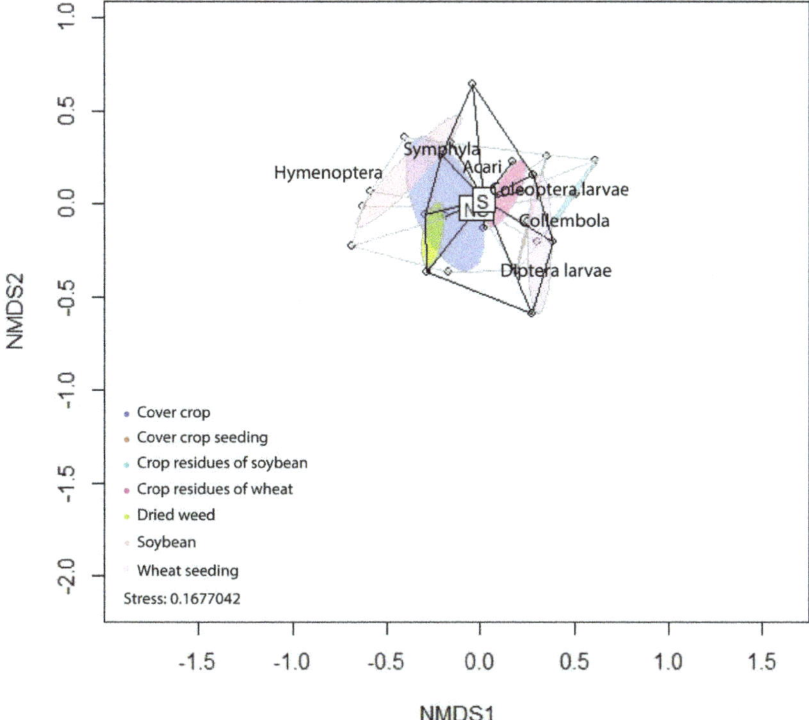

Figure 7. Bray–Curtis based NMDS plot of the arthropod community composition in the Cavallini farm. Points represent samples. "Spider" diagrams connect each point to the belonging irrigation system: NS: no sub-irrigation, in grey, and S: sub-irrigation, in black. Ellipses represent crop variables. To avoid overlapping, only the six more abundant groups were labelled.

3.4. Comparison between the Ruozzi and Gli Ulivi Farms

On the basis of the community data visualized through NMDS, Ruozzi and Gli Ulivi farms appeared to be more similar to each other than Cavallini (Figure 2), thereby allowing a comparison between their results. On both farms, total arthropod abundance appeared to be affected by crops and their interaction with management type, but not by management itself only. Moreover, a similar trend was observed in the differences found between 2014 and autumn 2015 in conservation management, and in autumn 2015 between conservation and conventional management, despite the different crops on the two farms (Figure 3A). On the other hand, the number of groups appeared to be a more sensitive variable on the Gli Ulivi farm, where it was affected by both management and crops and by their interaction, while only crop type showed an impact on this variable on Ruozzi farm (Figure 3B). On Ruozzi, crop type was the only factor affecting QBS-ar and the proportion of groups with maximum EMI, showing the same response for both variables (Figure 3C). On the other hand, on Gli Ulivi, QBS-ar did not appear to be significantly affected by the factors considered, while the proportion of groups with EMI 20 was influenced both by crops and their interaction with management type (Figure 3D). Changes in Acari and Collembola abundances followed roughly the same pattern on both farms, and generally resulted higher in wheat, mainly in conservation on Ruozzi, while the other most abundant groups did not appear to be linked to each other with a specific crop either (Figures 4 and 6). The abundances of Acari and Collembola were affected by the interaction between crops and by management type on Ruozzi and on Gli Ulivi, and on Ruozzi by crops and management type taken individually too (Figure 5A,B). High values were generally found in conservation management, in wheat on Ruozzi and in cover crop on Gli Ulivi. Hymenoptera were affected by crop types on both farms, but in no case by management (Figure 5C). Traditional and conservation management yielded similar results for the three groups in 2017 on both farms.

4. Discussion

Perturbations on agroecosystem are typically much greater than the ones occurring on other terrestrial ecosystems, particularly in the case of systems that are continuously cropped and subject to disturbance caused by cultivation and other agricultural practices [51]. Some studies have indicated land use, farming system (conventional or conservation), crop type and rotation in croplands, and other aspects related to management (e.g., use of pesticides, herbicides, fertilizers) as factors affecting soil fauna, both acting individually or interacting in agricultural landscapes [15,19,25,52,53]. By studying the effects of different soil managements and crops on soil fauna on three farms, our research partially confirms these observations: farm, intended as agricultural landscape, management, and crop type are patterns that drive differences in soil fauna assemblages; however, other factors like total abundance, diversity, or presence of adapted groups resulted more affected by crops than by management type.

Our results suggest that soil fauna variability, in terms of community composition, are largely related to not only crop type, but also farm characteristics. Indeed, the three farms differed in soil and climate conditions, and variations in soil type and properties could be important factors to determine soil-inhabiting communities according to other studies [30,54]. Moreover, we found that arthropod assemblages differ greatly depending on management and crop type. However, even if conservation management generally shows higher abundance and biodiversity when compared with conventional ones [55], in our study, we have found that they were not significantly affected by management type, unless the interaction with crop type is considered. This result agrees with the findings by Bedano [56], who observed no conclusive trends regarding no-till benefits when compared to reduced tillage or conventional tillage. Moreover, Tuck et al. [54] highlighted that differences could be hidden by the local management, in the sense that soil animals and chemicals can move through the landscape. Our results could therefore be affected by the proximity of the conventionally managed fields to the conservation ones, on both Ruozzi and Gli Ulivi farms. Furthermore, at the time of this study the conservation practices had only been introduced three years earlier, and their positive effects could be evident only after a longer time of application, especially on the arthropods which are more adapted to soil, and consequently more sensitive, as estimated by the QBS-ar index. Indeed, Fiorini et al. [55] highlighted the positive effects on QBS-ar index in a seven-year experimentation of no-till compared with conventional agriculture practices, possibly related to the enhancement of SOC sequestration potential, as well as a higher chance for edaphic fauna of developing morphological adaptation in soils subject to less disturbance.

Tuck et al. [54] highlighted significant differences in the effect of organic farming among crop types, mainly between cereals. We observed a similar trend on Ruozzi farm, where the soil arthropod abundance and diversity in corn differed from the one observed in wheat and wheat crop residues. We found generally lower values in corn for both abundance and taxa, as well as for soil quality, in terms of QBS-ar, and presence of groups more adapted to soil, notwithstanding a condition more favourable to soil fauna in conservation management, a result supported by Winter et al. [57]. Moreover, community assemblages in corn and wheat differed between managements, with a dissimilarity mainly due to Acari and Collembola, respectively. In conservation management, wheat showed the overall greatest abundance of soil arthropods, owing to the great number of Acari and, especially, Collembola and Hymenoptera, while the total number of arthropod groups were generally higher in crop residues of forage. The previous crop types in the field should be considered too, as noted by Cortet [15]. In this case, the influence of the previous crop is shown by the QBS-ar index, as well as by the proportion of taxa with EMI 20, which suggest that the field that supported the most adapted fauna was conservatively managed and had wheat residues. Conventional management showed a similar trend in bare soil. This condition can be explained by the lower anthropogenic activity related to the absence of a specific crop, followed by tillage as the only agricultural practice. Moreover, bare soil community assemblages were similar for more than 60% to those found in conservation management with crop residues of wheat. The effect of the previous crop appears to be emphasised on Gli Ulivi farm, too, where wheat after alfalfa crop supported a higher number of groups, along with Hymenoptera abundance, than wheat in 2017, perhaps a consequence of growing sorghum in the gap between cover crop and wheat. Indeed, in both managements, community assemblage too differed more than 40% between wheat after alfalfa and wheat in 2017, with Hymenoptera accounting for the greater difference. Generally, as on the Ruozzi farm, results on Gli Ulivi agree with findings by Rizk [58], showing that soil management techniques and crops that enhance diversity, as well as biological quality of the soil, involved conservation management, cover crops, and crop residues. On Gli Ulivi, the difference between conventional and conservation farming was evident in bare soil compared to cover crop, for fauna composition, number of taxa, and abundance, mainly of Collembola. This difference is probably due to the harrowing on bare soil in early November, practice that can drastically reduce diversity and activity of soil fauna. This was supported by Maraun et al. [24], who suggested that Collembola are sensitive to mechanical disturbances, even more than mites.

As reported by Menta and Remelli [59], some arthropod groups are widely used to detect soil quality and the effects of soil managements; among these arthropods, Collembola and Acari are the two most important groups in terms of abundance and species diversity [60], and subsequently the most investigated taxa. In our study Acari and Collembola were the groups that accounted for the greater dissimilarities, both within and between managements, generally followed by Hymenoptera. Comparing the abundances of Acari and Collembola on the two farms, they are generally higher on Ruozzi conservation wheat than on Gli Ulivi, where wheat appears to be a less favourable crop. Van de Bund [61] suggested that, even if crops showed a considerable influence on the fauna of mites and springtails, the preference for living under a special crop was not similar under different types of soil. However, another explanation could be the use of slurry on Ruozzi farm, which could increase the abundance of some tolerant arthropods like Collembola, while more sensitive ones could disappear [62,63]. In this case, Symphyla, Chilopoda, and Coleoptera larvae were not present in Ruozzi conservation wheat. Nevertheless, confirming Bund [61]'s observation that the abundances of these groups were much greater within the root system of plants than in bare soil, the overall distribution of Collembola and Acari in our study was similar on both Ruozzi and Gli Ulivi farms. On Cavallini their relative abundance is lower and a higher proportion of other groups, such as Symphyla and Diptera larvae, is observed, especially under sub-irrigation systems. Since Cavallini fields are conservatively managed, Acari and Collembola would be expected to be more abundant, even if the use of herbicides and fungicides on the farm, throughout the years, might have affected these groups negatively, especially herbivore and fungivore collembolans. In those fields, where the difference in management system was based on the presence of sub-irrigation, crop type influenced fauna abundance and diversity, so that different groups were advantaged by different crops. Moreover, both the irrigation system and crop type influenced soil fauna assemblages. Under this conservation, management differences were driven by much more groups than in Ruozzi and Gli Ulivi farms, with some generally minor groups sometimes contributing more to the overall dissimilarity than Acari and Collembola, such as Hemiptera, that appeared particularly linked to soybean. On the other hand, decomposers generally prefer crop residues [64]. This is the case of Diptera larvae, widespread in crop residues and cover crop seedings, in the latter case probably enhanced by the lack of predators such as Araneae. Other taxa, such as Symphyla, generally take advantage of the interaction between crop and sub-irrigation system, together with conservation management. Indeed, Peachey et al. [65] observed that the number of Symphyla, which generally consume germinating seeds, plant roots, and plant parts in contact with the soil, may increase as a consequence of reduced tillage, notwithstanding that cover cropping seems to be the most powerful factor.

5. Conclusions

The aim of this study was to evaluate the effects of management systems and crop types on soil fauna. Considering both factors, crop type seemed to have a greater effect on the arthropod community, although the conservation system generally provides better conditions for soil fauna, often interacting with crop. Indeed, biodiversity, in terms of soil arthropod abundance and number of groups, and soil quality index, in terms of number of arthropods well adapted to soil, were higher in both conventionally and conservatively managed fields when less impacted by anthropogenic practices, such as cover crops and crop residues. However, arthropod assemblages respond to soil practices differently, thereby highlighting different sensitivity to soil agricultural management and crops, with Acari, Collembola, and Hymenoptera accounting for the major dissimilarities, as well as abundance. Further studies are needed to clarify the effects of different agricultural management on soil faunal dynamics in the era of agricultural sustainable intensification.

Supplementary Materials: The following are available online at http://www.mdpi.com/2073-4395/10/7/982/s1, Scheme S1: Farm management and crop type in the sampling period. Under each farm management, the land use in the study period was indicated. A: Ruozzi farm; B: Gli Ulivi farm; C: Cavallini farm. Table S1: Analytical data of the three farms soils. Table S2: Taxa abundance (ind./m$_2$) ± standard error for every crop type within the agricultural system in the three farms. A: Ruozzi farm; B: Gli Ulivi farm; C: Cavallini farm. Table S3: Results of SIMPER analysis in: A: Ruozzi farm; B: Gli Ulivi farm; C: Cavallini farm. Most influential arthropod groups accounting for a cumulative dissimilarity within (×1) and between (×2) management of 70% are shown.

Author Contributions: Conceptualization, C.M., F.S., and S.R.; methodology, C.M., F.S., and S.R.; validation, C.M., F.S., and S.R.; formal analysis, S.R.; investigation, C.M., F.S., and S.R.; data curation, C.M., F.D.C., C.L.F., F.S., and S.R.; writing—original draft preparation, C.M., F.D.C., C.L.F., F.S., and S.R.; writing—review and editing, C.M. and S.R.; visualization, C.M. and S.R.; supervision, C.M. and F.S.; project administration, C.M.; funding acquisition, C.M. All authors have read and agreed to the published version of the manuscript.

Funding: This research was funded by the Emilia-Romagna Region, grant number CIG ZE111C629A and CIG ZE01AA07FD.

Acknowledgments: We would like to thank the farmers of the Ruozzi, Gli Ulivi, and Cavallini farms for their hospitality during the soil sampling phase.

Conflicts of Interest: The authors declare no conflict of interest.

References

1. Shukla, P.R.; Skea, J.; Calvo Buendía, E.; Masson-Delmotte, V.; Pörtner, H.O.; Roberts, D.C.; Zhai, P.; Slade, R.; Connors, S.; van Diemen, R.; et al. *Climate Change and Land: An IPCC Special Report on Climate Change, Desertification, Land Degradation, Sustainable Land Management, Food Security and Greenhouse Gas Fluxes in Terrestrial Ecosystems*; IPCC: Geneva, Switzerland, 2019.

2. Barnes, A.D.; Allen, K.; Kreft, H.; Corre, M.D.; Jochum, M.; Veldkamp, E.; Clough, Y.; Daniel, R.; Darras, K.; Denmead, L.H.; et al. Direct and cascading impacts of tropical land-use change on multi-trophic biodiversity. *Nat. Ecol. Evol.* **2017**, *1*, 1511–1519. [CrossRef] [PubMed]

3. Kibblewhite, M.G.; Ritz, K.; Swift, M.J. Soil health in agricultural systems. *Philos. Trans. R. Soc. B Biol. Sci.* **2008**, *363*, 685–701. [CrossRef] [PubMed]

4. Smith, P.; Cotrufo, M.F.; Rumpel, C.; Paustian, K.; Kuikman, P.J.; Elliott, J.A.; McDowell, R.; Griffiths, R.I.; Asakawa, S.; Bustamante, M.; et al. Biogeochemical cycles and biodiversity as key drivers of ecosystem services provided by soils. *SOIL* **2015**, *1*, 665–685. [CrossRef]

5. Cates, A.M.; Ruark, M.D.; Hedtcke, J.L.; Posner, J.L. Long-term tillage, rotation and perennialization effects on particulate and aggregate soil organic matter. *Soil Tillage Res.* **2016**, *155*, 371–380. [CrossRef]

6. Gardi, C.; Montanarella, L.; Arrouays, D.; Bispo, A.; Lemanceau, P.; Jolivet, C.; Mulder, C.; Ranjard, L.; Römbke, J.; Rutgers, M.; et al. Soil biodiversity monitoring in Europe: Ongoing activities and challenges. *Eur. J. Soil Sci.* **2009**, *60*, 807–819. [CrossRef]

7. Tabaglio, V.; Gavazzi, C.; Menta, C. Physico-chemical indicators and microarthropod communities as influenced by no-till, conventional tillage and nitrogen fertilisation after four years of continuous maize. *Soil Tillage Res.* **2009**, *105*, 135–142. [CrossRef]

8. Magro, S.; Gutiérrez-López, M.; Casado, M.A.; Jiménez, M.D.; Trigo, D.; Mola, I.; Balaguer, L. Soil functionality at the roadside: Zooming in on a microarthropod community in an anthropogenic soil. *Ecol. Eng.* **2013**, *60*, 81–87. [CrossRef]

9. Vignozzi, N.; Agnelli, A.E.; Brandi, G.; Gagnarli, E.; Goggioli, D.; Lagomarsino, A.; Pellegrini, S.; Simoncini, S.; Simoni, S.; Valboa, G.; et al. Soil ecosystem functions in a high-density olive orchard managed by different soil conservation practices. *Appl. Soil Ecol.* **2019**, *134*, 64–76. [CrossRef]

10. Delgado-Baquerizo, M.; Reich, P.B.; Trivedi, C.; Eldridge, D.J.; Abades, S.; Alfaro, F.D.; Bastida, F.; Berhe, A.A.; Cutler, N.A.; Gallardo, A.; et al. Multiple elements of soil biodiversity drive ecosystem functions across biomes. *Nat. Ecol. Evol.* **2020**, *4*, 210–220. [CrossRef]

11. Lavelle, P.; Decaëns, T.; Aubert, M.; Barot, S.; Blouin, M.; Bureau, F.; Margerie, P.; Mora, P.; Rossi, J.P. Soil invertebrates and ecosystem services. *Eur. J. Soil Biol.* **2006**, *42*. [CrossRef]

12. Mulder, C.; Boit, A.; Bonkowski, M.; De Ruiter, P.C.; Mancinelli, G.; Van der Heijden, M.G.A.; Van Wijnen, H.J.; Vonk, J.A.; Rutgers, M. A Belowground Perspective on Dutch Agroecosystems: How Soil Organisms Interact to Support Ecosystem Services. In *Advances in Ecological Research*; Academic Press: Cambridge, MA, USA, 2011; Volume 44, pp. 277–357. [CrossRef]

13. Cole, L.; Buckland, S.M.; Bardgett, R.D. Influence of disturbance and nitrogen addition on plant and soil animal diversity in grassland. *Soil Biol. Biochem.* **2008**, *40*, 505–514. [CrossRef]

14. Bardgett, R.D.; Cook, R. Functional aspects of soil animal diversity in agricultural grasslands. *Appl. Soil Ecol.* **1998**, *10*, 263–276. [CrossRef]

15. Cortet, J.; Ronce, D.; Poinsot-Balaguer, N.; Beaufreton, C.; Chabert, A.; Viaux, P.; Cancela De Fonseca, J.P. Impacts of different agricultural practices on the biodiversity of microarthropod communities in arable crop systems. *Eur. J. Soil Biol.* **2002**, *38*, 239–244. [CrossRef]

16. Dubie, T.R.; Greenwood, C.M.; Godsey, C.; Payton, M.E. Effects of Tillage on Soil Microarthropods in Winter Wheat. *Southwest. Entomol.* **2011**, *36*, 11–20. [CrossRef]

17. Rodríguez, E.; Fernández-Anero, F.J.; Ruiz, P.; Campos, M. Soil arthropod abundance under conventional and no tillage in a Mediterranean climate. *Soil Tillage Res.* **2006**, *85*, 229–233. [CrossRef]

18. Stinner, B. Arthropods and Other Invertebrates In Conservation-Tillage Agriculture. *Annu. Rev. Entomol.* **1990**, *35*, 299–318. [CrossRef]

19. House, G.J.; Parmelee, R.W. Comparison of soil arthropods and earthworms from conventional and no-tillage agroecosystems. *Soil Tillage Res.* **1985**, *5*, 351–360. [CrossRef]

20. Crossley, D.A.; Mueller, B.R.; Perdue, J.C. Biodiversity of microarthropods in agricultural soils: Relations to processes. *Agric. Ecosyst. Environ.* **1992**, *40*, 37–46. [CrossRef]

21. Hendrix, P.F.; Parmelee, R.W.; Crossley, D.A.; Coleman, D.C.; Odum, E.P.; Groffman, P.M. Detritus Food Webs in Conventional and No-Tillage Agroecosystems. *Bioscience* **1986**, *36*, 374–380. [CrossRef]

22. Neher, D.; Barbercheck, M. Diversity and Function of Soil Mesofauna. In *Biodiversity in Agroecosystems*; Williams Collins, W., Qualset, C.O., Eds.; CRC Press: Boca Raton, FL, USA, 1998; pp. 27–47. [CrossRef]

23. Filser, J.; Mebes, K.H.; Winter, K.; Lang, A.; Kampichler, C. Long-term dynamics and interrelationships of soil Collembola and microorganisms in an arable landscape following land use change. *Geoderma* **2002**, *105*, 201–221. [CrossRef]

24. Maraun, M.; Salamon, J.A.; Schneider, K.; Schaefer, M.; Scheu, S. Oribatid mite and collembolan diversity, density and community structure in a moder beech forest (Fagus sylvatica): Effects of mechanical perturbations. *Soil Biol. Biochem.* **2003**, *35*, 1387–1394. [CrossRef]

25. Diekötter, T.; Wamser, S.; Wolters, V.; Birkhofer, K. Landscape and management effects on structure and function of soil arthropod communities in winter wheat. *Agric. Ecosyst. Environ.* **2010**, *137*, 108–112. [CrossRef]

26. Bedano, J.C.; Cantú, M.P.; Doucet, M.E. Soil springtails (Hexapoda: Collembola), symphylans and pauropods (Arthropoda: Myriapoda) under different management systems in agroecosystems of the subhumid Pampa (Argentina). *Eur. J. Soil Biol.* **2006**, *42*, 107–119. [CrossRef]

27. Palacios-Vargas, J.G. Protura y Diplura. In *Biodiversidad, Taxonomía Y Biogeografía de Artrópodos de México: Hacia Una Síntesis de su Conocimiento*; Llorente, J., González, E., Papayero, N., Eds.; UNAM: Mexico City, México, 2000; p. 275.

28. Meyer, M.; Ott, D.; Götze, P.; Koch, H.; Scherber, C. Crop identity and memory effects on aboveground arthropods in a long-term crop rotation experiment. *Ecol. Evol.* **2019**, *9*, 7307–7323. [CrossRef] [PubMed]

29. Jones, D.T.; Susilo, F.X.; Bignell, D.E.; Hardiwinoto, S.; Gillison, A.N.; Eggleton, P. Termite assemblage collapse along a land-use intensification gradient in lowland central Sumatra, Indonesia. *J. Appl. Ecol.* **2003**, *40*, 380–391. [CrossRef]

30. Curry, J.P. The Arthropod Fauna Associated with Cattle Manure Applied as Slurry to Grassland. *Proc. R. Irish Acad. Sect. B Biol. Geol. Chem. Sci.* **1979**, *79*, 15–27. [CrossRef]

31. Andrén, O.; Lagerlöf, J. Soil Fauna (Microarthropods, Enchytraeids, Nematodes) in Swedish Agricultural Cropping Systems. *Acta Agric. Scand.* **1983**, *33*, 33–52. [CrossRef]

32. Kautz, T.; López-Fando, C.; Ellmer, F. Abundance and biodiversity of soil microarthropods as influenced by different types of organic manure in a long-term field experiment in Central Spain. *Appl. Soil Ecol.* **2006**, *33*, 278–285. [CrossRef]

33. Cluzeau, D.; Guernion, M.; Chaussod, R.; Martin-Laurent, F.; Villenave, C.; Cortet, J.; Ruiz-Camacho, N.; Pernin, C.; Mateille, T.; Philippot, L.; et al. Integration of biodiversity in soil quality monitoring: Baselines for microbial and soil fauna parameters for different land-use types. *Eur. J. Soil Biol.* **2012**, *49*, 63–72. [CrossRef]

34. Gruss, I.; Twardowski, J.; Hurej, M. Influence of 90-Year Potato and Winter Rye Monocultures under Different Fertilisation on Soil Mites. *Plant Prot. Sci.* **2018**, *54*. [CrossRef]

35. Pollierer, M.M.; Langel, R.; Körner, C.; Maraun, M.; Scheu, S. The underestimated importance of belowground carbon input for forest soil animal food webs. *Ecol. Lett.* **2007**, *10*, 729–736. [CrossRef] [PubMed]

36. Weigel, H.J.; Pacholski, A.; Burkart, S.; Helal, M.; Heinemeyer, O.; Kleikamp, B.; Manderscheid, R.; Frühauf, C.; Hendrey, G.F.; Lewin, K.; et al. Carbon turnover in a crop rotation under free air CO_2 enrichment (FACE). *Pedosphere* **2005**, *15*, 728–738.

37. Sticht, C.; Schrader, S.; Giesemann, A.; Weigel, H.J. Atmospheric CO_2 enrichment induces life strategy- and species-specific responses of collembolans in the rhizosphere of sugar beet and winter wheat. *Soil Biol. Biochem.* **2008**, *40*, 1432–1445. [CrossRef]

38. Holland, J.M.; Luff, M.L. The effects of agricultural practices on Carabidae in temperate agroecosystems. *Integr. Pest Manag. Rev.* **2000**, *5*, 109–129. [CrossRef]

39. Parisi, V.; Menta, C.; Gardi, C.; Jacomini, C.; Mozzanica, E. Microarthropod communities as a tool to assess soil quality and biodiversity: A new approach in Italy. *Agric. Ecosyst. Environ.* **2005**, *105*, 323–333. [CrossRef]

40. *ArcGIS for Desktop*; Version 10.4.1; Environmental Systems Research Institute (ESRI): Redlands, CA, USA, 2010.

41. Google Earth Pro (v. 7.3.2.5776-64 Bits). (27 April 2019). Ruozzi Farm, Emilia-Romagna, Italy. 32T 639989.98 m E; 4952099.21 m N, Eye alt 962 m. DigitalGlobe 2020. Available online: http://www.earth.google.com (accessed on 28 May 2020).

42. Google Earth Pro (v. 7.3.2.5776-64 Bits). (27 April 2019). Gli Ulivi Farm, Emilia-Romagna, Italy. 32T 733954.26 m E; 4888816.93 m N, Eye alt 1.11 km. DigitalGlobe 2020. Available online: http://www.earth.google.com (accessed on 8 May 2020).

43. Google Earth Pro (v. 7.3.2.5776-64 Bits). (27 April 2019). Cavallini Farm, Emilia-Romagna, Italy. 32T 719093.79 m E; 4949785.82 m N, Eye alt 1.12 km. DigitalGlobe 2020. Available online: http://www.earth.google.com (accessed on 8 May 2020).

44. Ministro per le Politiche Agricole Metodi Ufficiali di Analisi Chimica del Suolo. Available online: https://www.gazzettaufficiale.it/eli/id/1999/10/21/099A8497/sg (accessed on 27 March 2020).

45. Menta, C.; Conti, F.D.; Pinto, S. Microarthropods biodiversity in natural, seminatural and cultivated soils—QBS-ar approach. *Appl. Soil Ecol.* **2018**, *123*, 740–743. [CrossRef]

46. Bates, D.; Mächler, M.; Bolker, B.M.; Walker, S.C. Fitting linear mixed-effects models using lme4. *J. Stat. Softw.* **2015**, *67*, 1–48. [CrossRef]

47. Lenth, R.V. Least-squares means: The R package lsmeans. *J. Stat. Softw.* **2016**, *69*, 1–33. [CrossRef]

48. Oksanen, J.; Blanchet, F.G.; Kindt, R.; Legendre, P.; O'Hara, R.B.; Simpson, G.L.; Solymos, P.; Stevens, M.H.H.; Wagner, H. Vegan: Community Ecology Package. Available online: http://R-Forge.R-project.org/projects/vegan/ (accessed on 4 April 2020).

49. McDonald, J. Tests for one measurement variable. In *Handbook of Biological Statistics*; Publishing, S.H., Ed.; Sparky House Publishing: Baltimore, MD, USA, 2014; pp. 140–145, ISBN 9780444535924.

50. R Core Team. *R: A Language and Environment for Statistical Computing*; Version 3.6.3; R Foundation for Statistical Computing: Vienna, Austria, 2020.

51. Whalen, J.K.; Hamel, C. Effects of key soil organisms on nutrient dynamics in temperate agroecosystems. *J. Crop Improv.* **2004**, *11*, 175–207. [CrossRef]

52. Dekkers, T.B.M.; van der Werff, P.A.; van Amelsvoort, P.A.M. Soil Collembola and Acari related to farming systems and crop rotations in organic farming. *Acta Zool. Fenn.* **1994**, *195*, 28–31.

53. Boutin, C.; Martin, P.A.; Baril, A. Arthropod diversity as affected by agricultural management (organic and conventional farming), plant species, and landscape context. *Écoscience* **2009**, *16*, 492–501. [CrossRef]

54. Tuck, S.L.; Winqvist, C.; Mota, F.; Ahnström, J.; Turnbull, L.A.; Bengtsson, J. Land-use intensity and the effects of organic farming on biodiversity: A hierarchical meta-analysis. *J. Appl. Ecol.* **2014**, *51*, 746–755. [CrossRef] [PubMed]

55. Fiorini, A.; Boselli, R.; Maris, S.C.; Santelli, S.; Perego, A.; Acutis, M.; Brenna, S.; Tabaglio, V. Soil type and cropping system as drivers of soil quality indicators response to no-till: A 7-year field study. *Appl. Soil Ecol.* **2020**, *155*, 103646, in press. [CrossRef]

56. Bedano, J.; Domínguez, A. Large-Scale Agricultural Management and Soil Meso- and Macrofauna Conservation in the Argentine Pampas. *Sustainability* **2016**, *8*, 653. [CrossRef]

57. Winter, J.P.; Voroney, R.P.; Ainsworth, D.A. Soil Microarthropods In Long-Term No-Tillage And Conventional Tillage Corn Production. *Can. J. Soil Sci.* **1990**, *70*, 641–653. [CrossRef]

58. Rizk, M.A.; Mikhail, W.Z.A. Impact of no-tillage agriculture on soil fauna diversity. *Zool. Middle East* **1999**, *18*, 113–120. [CrossRef]

59. Menta, C.; Remelli, S. Soil Health and Arthropods: From Complex System to Worthwhile Investigation. *Insects* **2020**, *11*, 54. [CrossRef] [PubMed]

60. George, P.B.L.; Keith, A.M.; Creer, S.; Barrett, G.L.; Lebron, I.; Emmett, B.A.; Robinson, D.A.; Jones, D.L. Evaluation of mesofauna communities as soil quality indicators in a national-level monitoring programme. *Soil Biol. Biochem.* **2017**, *115*, 537–546. [CrossRef]

61. van de Bund, C.F. Influence of crop and tillage on mites and springtails in arable soil. *Netherlands J. Agric. Sci.* **1970**, *18*, 308–314.

62. Lübben, B.; Larink, O. Influence of sewage sludge fertilization and heavy metal content on Collembola in ploughed soil. *Ökologie Nat. im Agrar.* **1990**, *19*, 310–315.

63. Lameed, G.A. *Biodiversity Conservation and Utilization in a Diverse World*; InTech: London, UK, 2012. [CrossRef]

64. House, G.J.; Alzugaray, M.D.R. Influence of Cover Cropping and No-Tillage Practices on Community Composition of Soil Arthropods in a North Carolina Agroecosystem. *Environ. Entomol.* **1989**, *18*, 302–307. [CrossRef]

65. Peachey, R.E.; Moldenke, A.; William, R.D.; Berry, R.; Ingham, E.; Groth, E. Effect of cover crops and tillage system on symphylan (Symphlya: *Scutigerella immaculata*, Newport) and *Pergamasus quisquiliarum* Canestrini (Acari: Mesostigmata) populations, and other soil organisms in agricultural soils. *Appl. Soil Ecol.* **2002**, *21*, 59–70. [CrossRef]

 *agronomy*

Article

The Diversification and Intensification of Crop Rotations under No-Till Promote Earthworm Abundance and Biomass

María Pía Rodríguez [1,2,*], Anahí Domínguez [1,2], Melisa Moreira Ferroni [1], Luis Gabriel Wall [2,3] and José Camilo Bedano [1,2]

[1] Research Group in Ecology of Terrestrial Ecosystems (GIEET), Institute of Soil Sciences, Biodiversity and Environment (ICBIA), National University of Río Cuarto, Ruta Nac. 36 - Km. 601, X5804BYA Río Cuarto, Argentina; adominguez@exa.unrc.edu.ar (A.D.); mmoreiraferroni@gmail.com (M.M.F.); jbedano@gmail.com (J.C.B.)

[2] CONICET, National Council for Scientific and Technical Research, Godoy Cruz 2290, C1425FQB CABA, Argentina; wall.luisgabriel@gmail.com

[3] Department of Science and Technology, National University of Quilmes, Roque Sáenz Peña 352, B1876BXD Bernal, Argentina

[*] Correspondence: mprodriguez@exa.unrc.edu.ar

Received: 22 May 2020; Accepted: 17 June 2020; Published: 27 June 2020

Abstract: The diversification and intensification of crop rotations (DICR) in no-till systems is a novel approach that aims to increase crop production, together with decreasing environmental impact. Our objective was to analyze the effect of different levels of DICR on the abundance, biomass, and species composition of earthworm communities in Argentinean Pampas. We studied three levels of DICR—typical rotation (TY), high intensification with grass (HG), and with legume (HL); along with three references—natural grassland (NG), pasture (PA), and an agricultural external reference (ER). The NG had the highest earthworm abundance. Among the DICR treatments, abundance and biomass were higher in HL than in HG and, in both, these were higher than in TY. The NG and PA had a distinctive taxonomic composition and higher species richness. Instead, the DICR treatments had a similar richness and species composition. Earthworm abundance and biomass were positively related to rotation intensity and legume proportion indices, carbon input, and particulate organic matter content. The application of DICR for four years, mainly with legumes, favors the development of earthworm populations. This means that a subtle change in management, as DICR, can have a positive impact on earthworms, and thus on earthworm-mediated ecosystem services, which are important for crop production.

Keywords: soil; soil properties; macrofauna; earthworms; biodiversity; sustainability; soil invertebrates; farming systems

---

1. Introduction

In the early 1990s in Argentina, genetically modified soybean cropping was approved. After this, soybean monocropping, a wide adoption of no-till and an expansion of the agricultural area in detriment of natural ecosystems occurred, and is still being carried out in the Pampas region. The current agricultural system that prevails in our country is based on simplified practices, with very low crop diversity, and it generates soils impoverished in structure and nutrients. Furthermore, it is a system that is highly dependent on chemical inputs and GMOs [1–3]. However, in the last years, no-till farmers attempted to improve the simplified, low rotation, or monocropping systems, with the inclusion of "good agricultural practices" (GAP), as an integral part of no-till, i.e., mixed crop rotation, cover crops, integrated pest-weed and disease management, nutrient recycling, and a rational use of agrochemicals [4–6]. The GAP promoted a higher particulate organic carbon (POC) content [7] and induced favorable structural features [8] in the crop soils of the study region. Recently, some producers began to explore a new no-till alternative, which implies the diversification and intensification of crop rotations (DICR). Through means of intensifying the rotation sequences by including a greater number of crops per unit of time, a more efficient and intensive use of environmental resources, such as water and solar radiation is achieved. This higher efficiency allows us to maintain or increase crop production per unit of time and area, in a less harmful way, and contributes to a higher C return to the soil [9–11]. Moreover, well-balanced sequences between grasses and legumes provide stubbles that increase the contribution of C and N to the soil and consequently the productivity of the next crops [11].

Soil fauna perform important functions, including soil structure improvement, nutrient cycling, and organic matter decomposition. These processes might become much more important in no-till systems where there is no mechanical loosening of soil or mixing of soil and residues [12]. Among soil fauna, earthworms are a fundamental component. They are considered "ecosystem engineers" for their ability to directly or indirectly transform the availability of resources for their own benefit and for other species [13]. Earthworms improve soil structure, renew the organic matter, participate in the cycling of nutrients, and modify the bacterial community [14–17], directly and indirectly favoring plant productivity [18]. In a recent meta-analysis, Van Groenigen et al. [19] estimated that the presence of earthworms in agroecosystems produced an average increase of 25% in crop yields and a 23% increase in the aerial plant biomass. In a previous study in the Pampas region, we demonstrated that in no-till systems with GAP, earthworms significantly contributed to C incorporation, via differential consumption of soils enriched in organic matter and the consequent enrichment of earthworm aggregates (in no-till, 100% more POC was found in earthworm aggregates than in the surrounding soil). Furthermore, we also demonstrated that they contributed to soil structure through the production of macroaggregates that are more stable to water disruption than those physically generated [20]. Therefore, the conservation of the earthworm community is a key aspect to develop strategies that aim to increase agricultural productivity in a more sustainable way.

Soil management practices affect earthworm populations by affecting the food supply, mulch protection, and the chemical and physical environment [12]. The beneficial effects of no-till on earthworm populations compared to conventional tillage is widely demonstrated [21–23]. Some authors also highlight the importance of incorporating cover crops and of diversifying rotations to increase earthworm biomass and abundance in agricultural systems (e.g., [18,24]). Although several studies were carried out on earthworm communities in the agricultural soils of the Pampas region [4,25–27], and the positive effect of the inclusion of GAP was demonstrated [6], there are no studies that evaluated the effect of the different levels of diversification and intensification of crop rotation in no-till, on earthworm communities. The DICR is a relatively recent management approach through which economic sustainability is being tested by some producers, and whose impact on environmental sustainability indicators is extremely necessary to be evaluated. Therefore, the objective of the present study was to analyze the effect of different levels of diversification and intensification of crop rotations on the abundance, biomass, and species composition of the earthworm community. For this, we studied three levels of DICR in no-till systems—(1) typical rotation, (2) high intensification with grass, and (3) high intensification with legume. In addition, two internal references systems, a natural grassland and a long-term pasture, and an external reference of no-till with low rotation were studied. We hypothesized that—(1) earthworm abundance and biomass will be higher in the natural grassland than in the agricultural sites, and among them, on the higher rotations with higher DICR levels, with the highest positive effect on legume than grass rotation; and (2) the natural grassland will have a higher species richness and different community composition, compared to the agricultural treatments, because they can harbor species that are highly sensitive to the disturbances produced by agriculture. Among agricultural treatments, the most intensified and diversified rotation will have a community composition more similar to that of the natural grassland than the lower rotation treatments, because they contributed to a more diverse and better quality food for earthworms.

2. Material and Methods

2.1. Study Area

A field experiment with different levels of DICR was carried out in four localities in the most productive area of the Pampas region of Argentina. Two localities were localized in the Buenos Aires province, close to the Ines Indart (34°23′48″ S, 60°32′29″ W; "La Matilde" farm) and Baradero (33°48′37″ S, 59°30′17″ W; "Las Matreras" farm) cities. The two others were localized in the Santa Fe province, near the Venado Tuerto (33°44′40″ S, 61°58′09″ W; "Carmen" farm) and Uranga (33°15′44″ S, 60°42′28″ W; "San Nicolas" farm) cities (Figure 1). Soils in the area were Mollisols (USDA Soil Taxonomy) or Phaeozem (World Reference Base for Soil Resources); in La Matilde, Las Matreras, and San Nicolas, the soils were Typic Argiudolls (silty clay loam), according to the USDA classification, with a well-developed illuvial horizon (Bt), while the Hapludolls soils dominate in Carmen, with a higher sand proportion [28,29]. Climate in the region was temperate sub-humid with a dry season in winter, and with a mean annual temperature at about 16–18 °C. The mean annual precipitation in the La Matilde, Las Matreras, and San Nicolas was about 950–1100 mm, while in Carmen, it was about 850–950 mm. The relief in the region was flat with a gentle slope, which was, in all sites, lower than 0.5%.

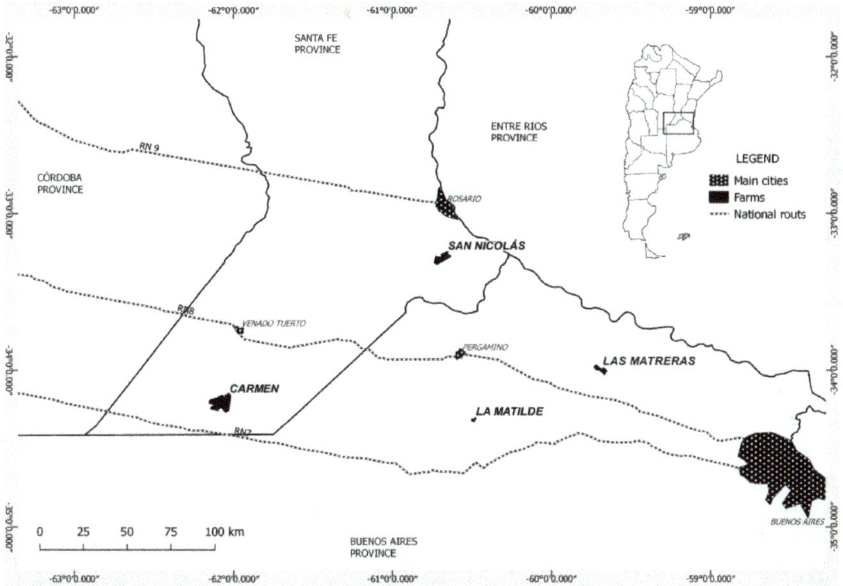

Figure 1. Study area in the Pampas region of Argentina. The farms are represented by black-filled figures, the main cities are represented by the dotted figures and the routes are shown by the dotted lines.

2.2. Experimental Design and Sampling

In 2011, farmers started the DICR field experiment in 4 localities. In each one, a large plot with a single and homogeneous land-use history was selected for performing the DIRC experiment. In the 10 years prior to the beginning of the experiment, every plot was managed under no-till, with soybean in the low-rotation scheme, and corn as the unique summer crop. Wheat was the only winter crop, although bare soil during winter was a frequent situation. In each locality, these plots were subdivided into four smaller plots, about 10 to 25 hectares for each one. In three of these plots, a crop rotation scheme of 3-year cycles with variations in crop intensity and diversity was established. The rotation scheme involved three DICR levels—TY (typical rotation), HG (high intensification with grass), and HL (high intensification with legumes) (Table 1). The fourth plot was cropped with a consociated pasture legume/grass (PA), which is considered to be an internal reference system. Additionally, two other references were selected—(1) natural grassland (NG), a site located in each farm with more than 30 years without agricultural intervention, with a mix of native and exotic grasses but without trees; and (2) an agricultural external reference (ER), an agricultural plot located near each farm, selected as representing the usual agricultural management of the region (no-till with low-crop rotation or soybean monocropping).

Table 1. Rotation scheme applied in each farm from the beginning of the essay. The sampling months are highlighted with an arrow. 1°: First sowing date (between October–December), 2°: second sowing date (from December) of the summer crops.

Farm	Plot	Agricultural year 2011–2012	Agricultural year 2012–2013	Agricultural year 2013–2014	Agricultural year 2014–2015	Agricultural year 2015–2016
Las Matreras	TY	Corn 1°	Soy 1°	Wheat — Soy 2°	Corn 1°	Soy 1°
Las Matreras	HL	Corn 2° — Pea	Soy 2° — Wheat	Corn 2°	Pea — Soy 2°	Soy 1°
Las Matreras	HG	Sorghum 1° — Wheat	Corn 2°	Sorghum 1° — Wheat	Corn 2°	Sorghum 1°
San Nicolas	TY	Soy 2°	Corn 1°	Soy 1°	Wheat — Soy 2°	Corn 1°
San Nicolas	HL	Soy 2° — Vetch	Corn 2°	Wheat — Soy 2°	Vetch — Corn 2°	Wheat — Soy 2°
San Nicolas	HG	Sorghum 2° — Wheat	Corn 2° — Barley	Sorghum 2°	Wheat — Corn 2°	Barley — Sorghum 2°
La Matilde	TY	Corn 2°	Sorghum 1° — Pea	Corn 2°	Sorghum 1°	Pea — Corn 2°
La Matilde	HL	Corn 2° — Wheat	Soy 2° — Pea	Corn 2° — Wheat	Soy 2°	Pea — Corn 2°
La Matilde	HG	Sorghum 2° — Vetch	Corn 2° — Wheat	Sorghum 2° — Vetch	Corn 2° — Wheat	Sorghum 2°
Carmen	TY		Wheat — Soy 2°	Corn 1°	Soy 1°	Wheat — Soy 2°
Carmen	HG		Wheat — Sorghum 2°	Barley — Corn 2°	Wheat — Sorghum 2°	Barley — Corn 2°
Carmen	HL		Wheat — Corn 2°	Wheat — Soy 2°	Barley — Corn 2°	Wheat — Soy 2°

TY—typical rotation, HG—high intensification with grass, HL—high intensification with legume, PA—pasture, NG—natural grassland, and ER—external reference. Soy —Soybean.

Once complete rotation cycle was attempted, we conducted our first sampling in May 2015 and a second (monitoring) sampling in May 2016. In 2015, 24 plots were sampled, consisting of 6 treatments (TY, HG, HL, NG, PA, and ER) in 4 localities (6 treatments × 4 localities), while in 2016, 12 plots were sampled since (i) the trial in Carmen farm was discontinued by the farmers and (ii) only three DICR levels, plus the PA as a reference were monitored (4 treatments × 3 localities). NG and ER were not sampled because both are expected to be more stable over time, given the absence of environmental or management changes; although this is also true for the PA, it was kept in the 2016 sampling, as an important reference for being part of the same initial large plot than DICR treatments.

Two rotation indices were used to characterize the DICR levels, this is, the occupation time by crops in rotations, expressed as crops per year of rotation [29,30]:

$$IRI\ (Intensification\ rotation\ index) = EPM/TDR,$$

where "EPM" are the days since crop emergence until physiological maturity, and "TDR" are the total days of rotation; and

$$ILI\ (Intensification\ legume\ index) = LEPM/TDR,$$

where "LEPM" are the days since emergence until physiological maturity of the legume crop.

Additionally, the average carbon input (CIn) by year for each rotation was calculated as:

$$CIn = \sum CIn_c\ (Bio \times 0.4 \times HC)/YR$$

where "$CInc$" is the humified carbon input by each crop in rotation, "Bio" is the total biomass contributed by the crop (aerial and root biomass), "0.4" is the C content of dry matter, "HC" is the humification coefficient of the crop, and "YR" refers to the years of the rotation [29,31].

2.3. Earthworms

Earthworms were sampled by the standardized sampling method in ISO [32]. In each sampling plot, five random sampling points were selected while avoiding the plot edges. In each point, a soil monolith of $25 \times 25 \times 20$ cm was extracted and split into layers—0 to 10 cm and 10 to 20 cm in depth. The earthworms were obtained carefully by hand-sorting the soil sample and then fixed in 96% alcohol. Once in the laboratory, earthworms were counted and weighed, and the adults were identified to the species level, using the taxonomic keys of Righi [33], Mischis and Moreno [34], Blakemore [35], and Momo and Falco [36]. Numbers and biomass of earthworms obtained from each monolith (0.0125 m^2) were expressed to 1 m^2. Biomass was not obtained in 2016 because juveniles were bred to maturity for taxonomic identification.

2.4. Soil Parameters

At each sampling point, soil samples were collected to determinate the following soil parameters:

Bulk density (BD) was determined by the cylinder method [37]. Soil samples were taken from 0–10 cm and 10–20 cm soil depth as duplicate undisturbed samples, using 100 cm^3 cylinders. Each sample was wet-weighted, dried in the oven for 24 h at 105 °C, and weighted again, to perform the calculation.

Particulate organic carbon (POC) after earthworm hand-sorting was carried out by collecting 100 g of soil from each monolith. The physical soil fractionation by particle size was conducted by the wet sieving method described in [38], obtaining two fractions: <53 µ and >53 µ. The determination of the CO content of each fraction was quantified by the Walkley & Black method [39].

Stubble biomass (Bio), i.e., the vegetable cover of each monolith was collected, dried in the oven at 40 °C, and weighted.

2.5. Statistical Analyses

The effect of the treatments on earthworm abundance was analyzed by generalized mixed linear models (GMLM). The models selected were those that presented the lowest AIC value [40] and a ratio (deviance)/(degrees of freedom) lower than 2.5 [41]. The model selected for the 2015 data was as follows—the treatment (TY, HG, HL, PA, NG, and ER) was considered as fixed factor; the locality (4), the depth (0–10 and 10–20 cm), and the sample (nested into treatment and locality) were the random factors. To analyze the effect of DICR treatments plus PA (reference) in both sampling years, a second model was performed by considering the abundance data of 2015 and 2016, where only the three DICR levels and the pastures were monitored. The selected model had the treatment (TY, HG, HL and PA) as a fixed factor, and the locality (except CA), the year, and the sample (nested into treatment and locality) as random factors. Due to data overdispersion, the abundance values were adjusted with a negative binomial distribution. The Di Rienzo, Guzman, and Casanoves (DGC) a posteriori test [42] was used to evaluate the significance of differences between treatments, when the p values were significant ($p < 0.01$). For earthworm biomass, the effect of the treatments was analyzed by the general mixed linear model and the model with the lowest AIC value was selected [43]. Prior to the analysis, data were transformed with base 10 logarithm (Log 10 (x + 1)) to fit the normal distribution. The *VarIdent* function was used to decrease the heterocedasticity of the data. In the selected model, the treatment (TY, HG, HL, PA, NG, and ER) was considered to be a fixed factor and the locality and the depth (0–10 and 10–20 cm) were the random factors. To assess the significant differences between management ($p < 0.01$), DGC was used as a posteriori test [42]. The species richness was analyzed by the general mixed linear model [43]. The treatment (TY, HG, HL, PA, NG, and ER) was considered as a fixed factor, while the locality was considered as a random factor. To assess the significant differences between management ($p < 0.01$), DGC was used as a posteriori test [42]. In both, biomass and richness models, the assumptions of variance homogeneity and normality were analyzed graphically, and in addition, normality was corroborated with the Shapiro-Wilks test.

To evaluate changes in the species composition of earthworm communities between the different treatments, principal component analysis (PCA) was performed. Prior to the analysis the abundance data were transformed according to Hellinger [44].

The relationships between earthworm abundance and biomass in 2015, with the soil and the rotation parameters, were analyzed by mixed models. Each soil (Bio, BD, and POC) and rotation (IRI, ILI, and CIn) parameter was considered as a fixed factor, and the locality was considered to be a random factor. The earthworm abundance data were analyzed by generalized mixed linear models with negative binomial distribution, while the log-transformed biomass data were analyzed by general mixed linear models. For each model, conditional R^2 was calculated using the "*r.squaredGLMM*" function (which describes the proportion of variance explained by both the fixed and random factors). The regressions of mixed models were plotted using the *VISREG* function [45].

All analyses were performed in R [46] and the Infostat [47] software.

3. Results

3.1. Earthworm Communities

The earthworm abundance in 2015 was affected by treatments ($p < 0.0001$) (Figure 2). The highest abundance was observed in NG, which presented more than twice the observed abundance in HL and ER. Among the DICR treatments, earthworm abundance was about twice in HL than in the HG and TY rotations, which had the lowest earthworm abundances. As well as earthworm abundance, the biomass in 2015 was also different between treatments ($p = 0.0034$) (Figure 3). The PA, NG, HL, and ER had the highest biomass values. Among the DICR treatments, the HL had the highest biomass, almost twice than in HG and TY, which had the lowest biomass of all treatments.

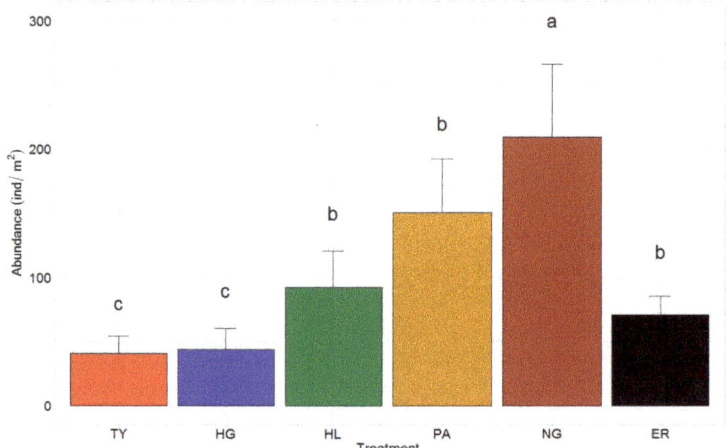

Figure 2. Earthworm abundance in the different treatments in 2015. Different letters indicate significant differences between treatments (DGC, $p < 0.05$). TY—typical rotation, HG—high intensification with grass, HL—high intensification with legume, PA—pasture, NG—natural grassland, ER—external reference, and AIC—1140.72.

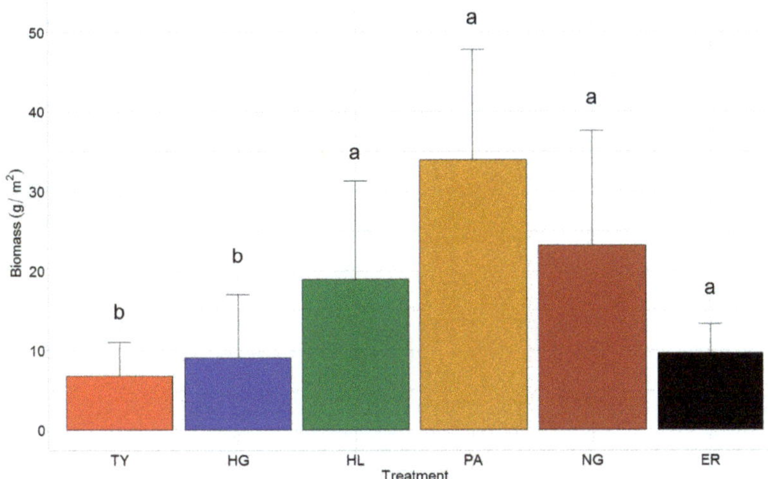

Figure 3. Earthworm biomass in the different treatments in 2015. Different letters indicate significant differences between treatments (DGC, $p < 0.05$). TY—typical rotation, HG—high intensification with grass, HL—high intensification with legume, PA—pasture, NG—natural grassland, ER—external reference, and AIC—11.72.

In the second abundance model, where the 2015 data of the DICR levels and the PA were analyzed together with the data from 2016 monitoring, the effect of treatments on earthworm abundance was significant ($p = 0.0102$); variance explained by year as a random factor was very low (0.11) suggesting DICR treatment as the main explaining factor. The PA, HL, and HG had greater abundances with respect to TY, confirming the observed differences in the first sampling year.

The species richness was significantly higher in NG (10 spp.) and in PA (7 spp.), than in ER (4 spp.), HL (2 spp.), and in both HG and TY (3 spp.).

Regarding species composition of earthworm communities in the NG, a distinctive species composition from the rest of the treatments was observed (PC1 50.4%), given by the presence of some exclusive species like *Metaphire californica*, *Amynthas gracilis*, *Glossodrilus parecis*, *Kenleenus armadas* and *Aporrectodea rosea* (Figure 4). The PA also presented a different species composition with respect to most of the treatments, which is mainly evident through axis 2 (PC2 23.9%). An association of HL, HG, and TY was observed, mainly due to the high abundance of *Aporrectodea caliginosa* and *Octolasion cyaneum* in the three treatments.

3.2. Earthworm Relationships with Soil Properties and Management Parameters

Earthworm abundance and biomass were positively related to intensification rotation index (IRI) (vs. abundance = R^2 0.7769, $p = 0.0012$; vs. biomass = R^2 0.7455, $p = 0.0059$) and ILI (vs. abundance = R^2 0.9329, $p = 3.95 \times 10^{-12}$; vs. biomass = R^2 0.7284, $p = 0.0109$) indices and to CIn (vs. abundance = R^2 0.7076, $p = 0.00562$; vs. biomass = R^2 0.7132, $p = 0.00728$) (Figure 5). In the case of IRI, the observed pattern was also associated with treatments (Figure 5a,b); the PA was associated to the highest index values and the highest values of abundance or biomass, the TY to the lowest and the HG and HL to the intermediate values of both index and earthworm abundance or biomass. Regarding ILI, a pattern of treatments association similar to that of IRI was observed, but the lowest values were observed in the HG and not in TY treatment (Figure 5c,d). In the case of CIn, there was only an association of the PA with the highest index values and highest earthworm abundance or biomass, but no pattern of treatment association was clear among the other treatments (Figure 5e,f).

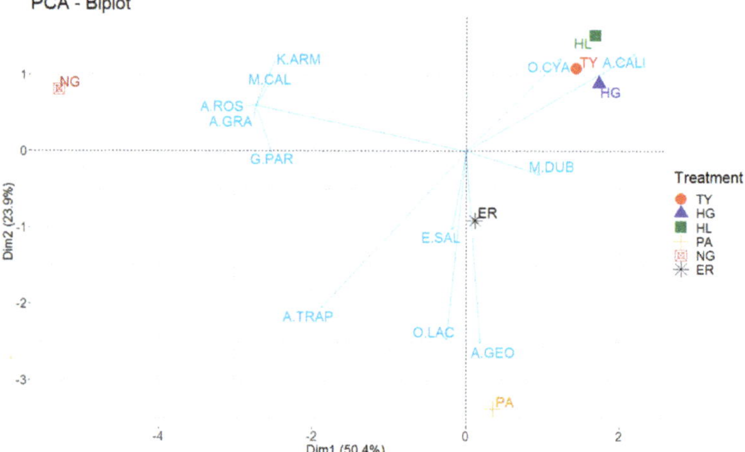

Figure 4. Principal component ordination diagram of Hellinger transformed earthworm community data showing the species vectors projected in the space formed by PCA axes 1 (50.4% of the variation) and 2 (23.9%); for all treatments in the four localities in 2015. Multiple arrows at the same point indicate an overlap of species. O.CYA (*Octolasion cyaneum*), A.CALI (*Aporrectodea caliginosa*), K.ARM (*Kenleenus armadas*), G.PAR (*Glossodrilus parecis*), O.LAC (*Octolasion lacteum*), M.DUB (*Microscolex dubius*), A.TRAP (*Aporrectodea trapezoides*), E.SAL (*Eukerria saltensis*), M.CAL (*Metaphire californica*), A.ROS (*Aporrectodea rosea*), A.GRA (*Amynthas gracilis*), and A.GEO (*Allolobophora georgii*). TY—typical rotation, HG—high intensification with grass, HL—high intensification with legume, PA—pasture, NG—natural grassland, and ER—external reference.

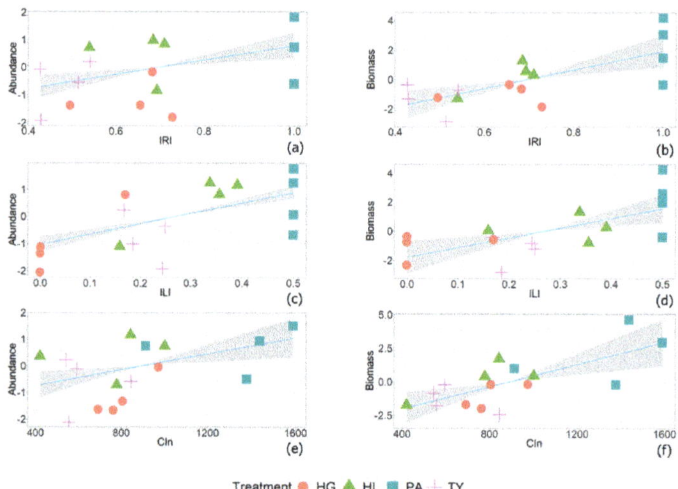

Figure 5. Relationship between intensification rotation index (IRI), intensification legume index (ILI), and Carbon input (CIn) with earthworm abundance and biomass, in the diversified and intensified rotations and pasture in 2015. (**a**) IRI vs. abundance, (**b**) IRI vs. biomass, (**c**) ILI vs. abundance, (**d**) ILI vs. biomass, (**e**) CIn vs. abundance, (**f**) CIn vs. biomass. Shading areas are the confidence intervals 95%. TY—typical rotation, HG—high intensification with grass, HL—high intensification with legume, and PA—pasture.

Regarding the relationships with soil parameters, earthworm abundance and biomass showed a positive marginally significant regression with POC content (vs. abundance, $R^2 = 0.6704$, $p = 0.0584$; vs. biomass $R^2 = 0.6902$, $p = 0.0481$) (Figure 6). The PA was associated with a higher POC and abundance and biomass values, while among DICR treatments, no clear pattern of treatment association was detected. There were no significant relationships between BD and Bio with earthworm abundance and biomass (BD vs. abundance $p = 0.165$; vs. biomass, $p = 0.654$), (Bio vs. abundance $p = 0.2614$; vs. biomass, $p = 0.3392$).

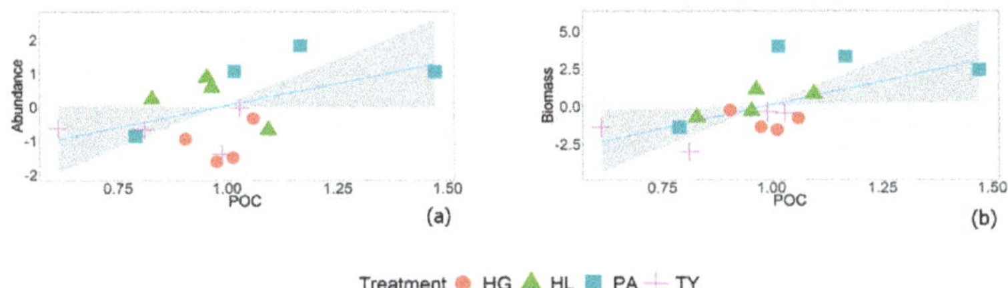

Figure 6. Relationship between earthworm abundance and biomass with soil particulate organic carbon content (POC) in 2015, in the DICR treatments, and PA. (**a**) POC vs. abundance, (**b**) POC vs. biomass. Shading areas are the confidence intervals of 95%. TY—typical rotation, HG—high intensification with grass, HL—high intensification with legume, and PA—pasture.

4. Discussion

The inclusion of good agricultural practices (GAP) in no-till systems, was shown to be positive on litter and soil fauna, however it was suggested that GAP as a management strategy, might be improved by increasing and diversifying crop-rotation intensity [6]. Therefore, in this contribution we studied the effect of three different levels of diversification and intensification of crop rotations under NT. The typical rotation, that could be considered as a similar intensification level as GAP, and two more intensified and diversified systems—high intensification with legumes and high intensification with grass.

4.1. DICR Effects on Earthworm Communities

As expected, the unmanaged grasslands supported the most abundant earthworm community. The abundance decline in the agricultural treatments was consistent with our hypothesis and with previous studies, either in other parts of the world [48] or in the study region [20,25,27,49,50]. The conversion of natural soils to agricultural ones implies deep, unfavorable changes in environmental conditions for earthworm communities [48,51]. In the Pampas region, the simplification of plant diversity, the soil compaction, and the strong dependence on herbicides, insecticides, and fungicides, are considered to be the main reasons of soil fauna reduction in agroecosystems [4].

As expected, the abundance of earthworms in the PA was significantly lower than in NG but higher than in the agricultural treatments. Moreover, both, the PA and NG showed the highest earthworm biomass. The positive effect of pastures on earthworms is likely related to the production of stubble of high nutritive quality, dead roots, and the presence of a permanent vegetation layer that protects earthworms from predation and extreme temperature fluctuations, which favors the development of earthworm populations, in a similar way to natural systems, reaching similar or even higher earthworm biomass [48,52,53]. Grass and legume consociate pastures (as implemented here) improve soil properties through nitrogen fixation by legumes and soil aggregation by grass root systems [54]. Furthermore, they offer a stubble that keeps quantity, quality, and continuity of food supply for earthworms through time [55,56].

Regarding the main objective of our study, our results showed that a four-year period (2011–2015) of intensification and diversification of crop rotations produced a clear positive effect on earthworm abundance and biomass, especially when the high DICR level included legumes. Moreover, this positive effect of DICR was sustained over time, according to the results obtained in the second sampling (2016). In the 2015 sampling, the rotation intensified with legumes (HL) showed a higher earthworm abundance and biomass than the typical rotation (TY), in agreement with our expectations of a higher positive effect of legume than grass rotation. A positive effect of legumes was previously observed on earthworm abundance and biomass [24,57]. Earthworm number and biomass were negatively correlated with the C/N ratio of the roots [57]; and, in short essays, larger weight gains in earthworms fed with residues with low C/N ratios was observed compared to high C/N diet [58]. Legume cover crops used in HL showed markedly lower C/N ratios (12 for pea [59] and 10 for vetch [60]) than the grass species used in HG (barley and wheat with C/N ratios of 109 and 102, respectively [61]). However, including data from the second sampling, the rotation with grasses promoted an earthworm abundance similar to the HL, both being higher than in TY. The difference in the response timing between HL and HG could be explained by the quality of the stubble provided, according to the time necessary for its decomposition and for its nutrients to be available. Crop species of low C/N ratio as legumes provide soil with residues of high nutritional quality and fast decomposition rate, being a fast-food source for earthworms in the short-term, while the opposite was observed for high C/N ratio residues, as grasses [62]. Thus, legume residues in HL provided a fast and better quality food resource for earthworms, promoting a greater abundance and biomass in the first year of sampling, while in HG due to the lower quality and slower decomposition rate of grass residues, the benefits on earthworm populations were observed when the data from the second year of sampling were included. At the same time, the TY rotation presented the lowest abundance and biomass values, demonstrating that the absence of winter cover crops and a high input of chemical herbicides, are unfavorable conditions for earthworm populations.

Regarding the external agricultural reference, in the 2015 sampling, it did not show statistical differences with HL, neither in abundance nor in biomass, although both parameters showed a trend to be higher in HL than in ER. The ER were fields external to the farm where the DICR treatments were performed, therefore, other unknown factors, mainly a different land use and management history, with respect to the field where the DICR assay was stablished, might have influenced this result. Species composition was influenced by biogeographical and historical factors, and thus different species react differently to management practices. Otherwise, this result highlights the importance of the positive effect of HL with respect to TY rotation, since the performance of the different treatments of the assay in the same field, guarantees that the observed differences were caused by the studied crop rotation changes.

In accordance with our hypothesis and with previous findings [63,64], the NG was characterized by an earthworm community different to the agricultural treatments, and also had the highest species richness. Habitat disturbance and physical and chemical alterations due to the change in land use differentially affect earthworm species, reducing the diversity of communities. It was suggested that this occurs because disturbances mainly affect the native species that are more susceptible to environmental change [63,65]. However, in our study the distinctive species composition in NG was given mainly by exotic but not by native earthworm species. This could be because the natural sites we sampled were small relicts that only partially conserve the characteristics from the original landscape and are often exposed to some degree of anthropogenic impact.

The PA had a community different to the DICR treatments and a higher richness with regards to them, similar to the NG. This was likely due to the favorable conditions that are generated in the pasture, where a permanent cover layer offered food and protection to earthworms [48,52,53], which favored a reconstitution of species number over time [48]. Although the species richness in the pasture was high ($r = 7$), close to that of the NG ($r = 10$), the composition of the community was different. As Decaens and Jimenez argued [48], when conditions provided by pastures were highly different from those of the initial natural vegetation, the recovery of the original community would be difficult to achieve. In this sense, the characteristics of consociate grass–legume pastures of 3 years sampled in this study were quite far from those of natural sites. However, its benefits in preserving a relatively diverse community are remarkable, considering that the original plot was the same than that for other DICR treatments.

Unlike what we hypothesized, there was no change in community composition between high intensified DICR treatments and TY. The richness in the three DICR systems was low (2 or 3 species), with a dominance of two exotic species—*Aporrectodea caliginosa* and *Octolasion cyaneum*. After four years of starting the DICR experiment, the intensification and diversification of crop rotations promoted earthworm abundance and biomass, but it did not change species composition. It is possible that, due to the low movement rate of earthworms, in general less than 10 m per year, and the limited availability of nearby natural patches that could act as species sources [64,66,67], four years are not enough to increase the richness and to cause changes in species composition. This is especially true when considering that all DICR treatments started from a single field with a homogeneous community. On the other hand, the successful adaptation of the species from the exotic family Lumbricidae, as *Aporrectodea caliginosa* and *Octolasion cyaneum*, in agricultural and cattle-raising fields in the region [68], might make it more difficult for the recolonization of native species in the short-term, due to competitive exclusion effects [64,69].

The ER had a community structure different to all other treatments, mainly characterized by higher *E. saltensis* abundances than the other systems. As we have said, ER system had a different location to the other treatments, and therefore its species composition result from different historical and geographical processes. The difference in earthworm communities among ER and the other systems, might also be related to the relative high abundances that we found, since different species might have different susceptibility to specific management practices.

4.2. Earthworm Relationships with Soil Properties and Management Parameters

The DICR implies an increase in the number of crops in the rotation, in order to increase the level of C input to the system and to improve the balance of C [29]. As expected, the regression models showed a positive relationship between earthworm abundance and biomass with both, IRI and ILI indices, and also with the CIn, showing that, for all analyzed samples, the DICR favored both earthworm abundance and biomass. The PA was generally associated with the highest values of intensification indices and CIn, and the highest values of abundance and biomass of earthworms. The already discussed benefits of the pastures for earthworms are highlighted by the regression analysis. Even more, because in this assay the PA was consociated with legumes and grasses, so in addition to providing a large biomass of litter as food for earthworms, it was of high quality. As the intensification indices decreased from the PA values, the abundance and biomass of earthworms also decreased. Regarding the three intensification treatments, both highly intensified rotations were different from the typical rotation in terms of IRI and both, earthworm abundance and biomass, followed that trend. In case of ILI, the clear response of earthworm abundance and biomass to the index increase confirmed the importance of legumes for earthworms, at least in the short-to-medium term, as in our study. Although the degree of intensification and the theoretical contribution of carbon were similar between the two highest intensified treatments, the rotation with legumes favored earthworms more than the rotation with grasses. As we have pointed out, legumes provide nutritious and high-quality stubble, being a fast food source for earthworms [55,62,70].

Among the soil parameters, the particulate organic carbon (POC) showed a positive relationship with earthworm abundance and biomass. This result agreed with previous studies, which recognize soil organic matter as a key factor for earthworm community development [25,71]. Moreover, the SOM is especially important for endogeic species, which live and feed within the soil [19] and that are the dominating species in the study region. There was no clear pattern relating the treatments with the results of the regression, which indicates that the relationship of POC with earthworms is independent of the treatments. There is a general link between the theoretical calculated C input with the POC levels measured in the soil, except for the typical rotation.

While the C input was higher in both highly intensified treatments than in the TY, this difference was not reflected in the soil POC content. It was likely that this four-year period of DICR was not enough to produce changes in SOM, even in a relatively rapid response parameter as POC.

5. Conclusions

In the present study, the diversification and intensification of crop rotations had a positive effect on both the abundance and the biomass of earthworms, mainly in the rotation that was highly intensified with legumes. We consider the magnitude of the effect to be compelling because of the short-term of the DICR experiment (4 years), and fundamentally because the differences among DICR treatments were relatively minor, compared to what is usually studied in agricultural systems (for example, monoculture vs. rotation, no-tillage vs. plow tillage). The greater input of high-quality trophic resources and the all-year growing roots, promotes the reproduction (more abundance) and growth (biomass) of the earthworms. However, the community structure did not change. For such a change, more time is needed, mainly because the earthworms have a limited migration capacity, moving slowly from one plot to another. In addition, a change in species composition depends on the landscape characteristics, such as the proximity of the cultivated plot to areas with greater species richness.

Overall, our results highlight the earthworm sensitivity to subtle changes in agricultural management and their importance as indicators. Moreover, in this region, earthworms are key drivers of C incorporation and the soil-structure maintenance processes [20]. This means that farmers' decisions (in this case applying DICR) are able to favor earthworm populations, and therefore to improve ecosystem services that are important for crop production. Then, we suggest that the earthworms should be considered when making decisions about agricultural managements.

Author Contributions: M.P.R., A.D., L.G.W., and J.C.B. conceived the idea and the experimental design, which support the manuscript, conducted the fieldwork, and laboratory determinations. M.M.F. contributed to the field and laboratory work. M.P.R., A.D., and J.C.B. conducted the statistical analysis and wrote the draft. M.M.F. and L.G.W. critically reviewed and contributed to the final manuscript. All authors approved the final article. All authors have read and agreed to the published version of the manuscript.

Funding: This study was funded by the project PICT 0851/06 of the Argentinean National Agency for Scientific and Technological Promotion (ANPCyT), the project SPOTT-Chacra Pergamino (UNQui-UNRC), and the SECyT (National University of Río Cuarto). A.D., L.G.W., and J.C.B. are members of the Argentinean National Council for Scientific and Technical Research (CONICET). M.P.R. is a fellow from CONICET.

Acknowledgments: We acknowledge the support provided by Belén Agosti in field work and in the provision and discussion of management data. We are also grateful to the farmers from the Regional Pergamino-Colón (AAPRESID) for their collaboration.

Conflicts of Interest: The authors declare no conflict of interest.

References

1. Pengue, W.A. Producción agroexportadora e (in) seguridad alimentaria: El caso de la soja en Argentina. *Revibec* **2004**, *1*, 46–55.
2. Aizen, M.A.; Garibaldi, L.A.; Dondo, M. Expansión de la soja y diversidad de la agricultura argentina. *Ecol. Austral* **2009**, *19*, 45–54.
3. Pengue, W.A. Cambios y Escenarios en la Agricultura Argentina del Siglo XXI. Available online: http://www.idaes.edu.ar/pdf_papeles/PENGUE_Agricultura%20Transformaciones%20Recursos%20y%20Escenarios%20en%20la%20Argentina%20FINAL%20ver%20SocialesBoll.pdf (accessed on 6 May 2020).
4. Bedano, J.C.; Domínguez, A. Large-scale agricultural management and soil meso-and macrofauna conservation in the Argentine Pampas. *Sustainability* **2016**, *8*, 653. [CrossRef]
5. Poisot, A.; Speedy, A.; Kueneman, E. Good Agricultural Practices—A working concept. In Proceedings of the FAO Internal Workshop on Good Agricultural Practices, Rome, Italy, 27–29 October 2004.
6. Bedano, J.C.; Domínguez, A.; Arolfo, R.; Wall, L.G. Effect of Good Agricultural Practices under no-till on litter and soil invertebrates in areas with different soil types. *Soil Till. Res.* **2016**, *158*, 100–109. [CrossRef]

7. Duval, M.E.; Galantini, J.A.; Iglesias, J.O.; Canelo, S.; Martinez, J.M.; Wall, L. Analysis of organic fractions as indicators of soil quality under natural and cultivated systems. *Soil Till. Res.* **2013**, *131*, 11–19. [CrossRef]

8. Kraemer, F.B.; Soria, M.A.; Castiglioni, M.G.; Duval, M.; Galantini, J.; Morrás, H. Morpho-structural evaluation of various soils subjected to different use intensity under no-tillage. *Soil Till. Res.* **2017**, *169*, 124–137. [CrossRef]

9. Caviglia, O.P.; Andrade, F.H. Sustainable intensification of agriculture in the Argentinean Pampas: Capture and use efficiency of environmental resources. *Am. J. Plant Sci. Biotechnol.* **2010**, *3*, 1–8.

10. Caviglia, O.P.; Sadras, V.O.; Andrade, F.H. Intensification of agriculture in the south-eastern Pampas: I. Capture and efficiency in the use of water and radiation in double-cropped wheat–soybean. *Field Crop. Res.* **2004**, *87*, 117–129. [CrossRef]

11. Andrade, J.F.; Poggio, S.L.; Ermacora, M.; Satorre, E.H. Land use intensification in the Rolling Pampa, Argentina: Diversifying crop sequences to increase yields and resource use. *Eur. J. Agron.* **2017**, *82*, 1–10. [CrossRef]

12. Kladivko, E.J. Tillage systems and soil ecology. *Soil Till. Res.* **2001**, *61*, 61–76. [CrossRef]

13. Jones, C.G.; Lawton, J.H.; Shachak, M. Organisms as ecosystem engineers. In *Ecosystem Management*; Samson, F.B., Knopf, F.L., Eds.; Springer: New York, NY, USA, 1994; pp. 130–147, ISBN 978-1-4612-4018-1.

14. Lavelle, P. Faunal activities and soil processes: Adaptive strategies that determine ecosystem function. *Adv. Ecol. Res.* **1997**, *27*, 93–132. [CrossRef]

15. Lavelle, P.; Spain, A.V. *Soil Ecology*; Springer: Dordrecht, The Netherlands, 2001; p. 619, ISBN-13 978-1-4020-0490-2 (PB).

16. Brown, G.G.; Doube, B.M. Functional interactions between earthworms, microorganisms, organic matter and plants. In *Earthworm Ecology*, 2nd ed.; Edwards, C.A., Ed.; CRC Press: Boca Raton, FL, USA, 2004; pp. 213–240, ISBN 0-8493-1819-X.

17. Brussaard, L. Ecosystem services provided by the soil biota. In *Soil Ecology and Ecosystem Services*; Wall, D.H., Ed.; Oxford University Press: Oxford, UK, 2012; pp. 45–58, ISBN 978-0-19-957592-3 (hbk.).

18. Bertrand, M.; Barot, S.; Blouin, M.; Whalen, J.; de Oliveira, T.; Roger-Estrade, J. Earthworm services for cropping systems. A review. *Agron. Sustain. Dev.* **2015**, *35*, 553–567. [CrossRef]

19. Van Groenigen, J.W.; Lubbers, I.M.; Vos, H.M.; Brown, G.G.; De Deyn, G.B.; Van Groenigen, K.J. Earthworms increase plant production: A meta-analysis. *Sci. Rep.* **2014**, *4*, 6365. [CrossRef]

20. Bedano, J.C.; Vaquero, F.; Domínguez, A.; Rodríguez, M.P.; Wall, L.; Lavelle, P. Earthworms contribute to ecosystem process in no-till systems with high crop rotation intensity in Argentina. *Acta Oecol.* **2019**, *98*, 14–24. [CrossRef]

21. Brown, G.G.; Benito, N.P.; Pasini, A.; Sautter, K.D.; de F Guimarães, M.; Torres, E. No-tillage greatly increases earthworm populations in Paraná state, Brazil: The 7th international symposium on earthworm ecology·Cardiff·Wales 2002. *Pedobiologia* **2003**, *47*, 764–771. [CrossRef]

22. Pelosi, C.; Pey, B.; Hedde, M.; Caro, G.; Capowiez, Y.; Guernion, M.; Peigné, J.; Piron, D.; Bertrand, M.; Cluzeau, D. Reducing tillage in cultivated fields increases earthworm functional diversity. *Appl. Soil Ecol.* **2014**, *83*, 79–87. [CrossRef]

23. Crittenden, S.J.; Huerta, E.; De Goede, R.G.M.; Pulleman, M.M. Earthworm assemblages as affected by field margin strips and tillage intensity: An on-farm approach. *Eur. J. Soil Biol.* **2015**, *66*, 49–56. [CrossRef]

24. Roarty, S.; Hackett, R.A.; Schmidt, O. Earthworm populations in twelve cover crop and weed management combinations. *Appl. Soil Ecol.* **2017**, *114*, 142–151. [CrossRef]

25. Domínguez, A.; Bedano, J.C.; Becker, A.R. Negative effects of no-till on soil macrofauna and litter decomposition in Argentina as compared with natural grasslands. *Soil Till. Res.* **2010**, *110*, 51–59. [CrossRef]

26. Falco, L.B.; Sandler, R.; Momo, F.; Di Ciocco, C.; Saravia, L.; Coviella, C. Earthworm assemblages in different intensity of agricultural uses and their relation to edaphic variables. *PeerJ* **2015**, *3*, e979. [CrossRef]

27. Domínguez, A.; Bedano, J.C. Earthworm and enchytraeid co-occurrence pattern in organic and conventional farming: Consequences for ecosystem engineering. *Soil Sci.* **2016**, *181*, 148–156. [CrossRef]

28. Soil Survey Staff. *Keys to Soil Taxonomy*, 12th ed.; USDA-Natural Resources Conservation Service: Washington, DC, USA, 2014; p. 372.

29. Agosti, M.B.; Madias, A.; Gil, R. *Informe Anual de Resultados Campaña 2015–2016 Chacra Pergamino*; INTA and Sistema Chacras-Aapresid: Pergamino, Argentina, 2016.

30. Farahani, H.J.; Peterson, G.A.; Westfall, D.G. Dry land cropping intensification: A fundamental solution to efficient use of precipitation. *Adv. Agron.* **1998**, *64*, 197–223.

31. Andriulo, A.; Mary, B.; Guerif, J. Modelling soil carbon dynamics with various cropping sequences on the rolling pampas. *Agronomie* **1999**, *19*, 365–377. [CrossRef]

32. International Organization for Standardization (ISO). *Soil Quality—Sampling of Soil Invertebrates—Part 1: Hand-sorting and Formalin Extraction of Earthworms. ISO 23611-1:2006*; International Organization for Standardization: Geneva, Switzerland, 2006.

33. Righi, G. Introducción al estudio de los Oligoquetos Megadrilos de la Provincia de Santa Fe. *Rev. As. Cs. Nat. Litoral* **1979**, *10*, 89–155.

34. Mischis, C.; Moreno, A.G. *Taxonomía de Oligoquetos: Criterios y Metodologías*; Curso de Postgrado, Universidad Nacional de Córdoba: Córdoba, Argentina, 1999.

35. Blakemore, R.J. *Cosmopolitan Earthworms: An Eco-Taxonomic Guide to the Peregrine Species of the World*; Verm Ecology: Kippax, ACT, Australia, 2002; p. 586.

36. Momo, F.R.; Falco, L.B. Las lombrices de tierra. In *Biología y Ecología de la Fauna del Suelo*; Momo, F.R., Falco, L.B., Eds.; Imago Mundi: Longchamps, Argentina, 2010; pp. 141–160, ISBN 9789507930942.

37. Blake, G.R.; Hartge, K.H. Bulk Density. In *Methods of soil Analisys*, 2nd ed.; Klute, A., Ed.; American Society of Agronomy Madison: Wisconsin, WI, USA, 1986; pp. 363–375.

38. Galantini, J.A. Separación y análisis de las fracciones orgánicas. In *Información y Tecnología en los Laboratorios de Suelos para el Desarrollo Agropecuario Sostenible*; Marbán., L., Ratto, S.E., Eds.; Asociación Argentina de Ciencia del Suelo: Buenos Aires, Argentina, 2005; pp. 103–114.

39. Jackson, M.L. *Análisis Químico de Suelos*, 1st ed.; Omega SA: Barcelona, España, 1976; p. 662, ISBN-13 9788428202619.

40. Burnham, K.P.; Anderson, D.R. *Model Selection and Multimodel Inference: A Practical Information-Theoretic Approach*, 2nd ed.; Springer: New York, NY, USA, 2002; p. 485, ISBN 0-387-95364-7.

41. Di Rienzo, J.A.; Macchiavelli, R.; Casanoves, F. *Modelos lineales Generalizados Mixtos Aplicaciones en InfoStat*, 1st special ed.; Julio Alejandro Di Rienzo: Córdoba, Argentina, 2017; p. 101, ISBN 978-987-42-4985-2.

42. Di Rienzo, J.A.; Guzmán, A.W.; Casanoves, F. A multiple-comparisons method based on the distribution of the root node distance of a binary tree. *J. Agric. Biol. Environ. Stat.* **2002**, *7*, 129–142. [CrossRef]

43. Zuur, A.; Ieno, E.N.; Walker, N.; Saveliev, A.A.; Smith, G.M. *Mixed Effects Models and Extensions in Ecology with R*; Springer Science & Business Media: New York, NY, USA, 2009; p. 573, ISBN 978-0-387-87458-6.

44. Legendre, P.; Legendre, L. *Numerical Ecology*, 2nd ed.; Elsevier Science: Amsterdam, The Netherlands; New York, NY, USA, 1998; p. 852.

45. Breheny, P.; Burchett, W. Visualization of regression models using visreg. *R J.* **2017**, *9*, 56–71. [CrossRef]

46. R Core Team. *R: A Language and Environment for Statistical Computing*; R Foundation for Statistical Computing: Vienna, Austria, 2017. Available online: https://www.R-project.org/ (accessed on 6 May 2020).

47. Di Rienzo, J.A.; Casanoves, F.; Balzarini, M.G.; Gonzalez, L.; Tablada, M.; Robledo, C.W. InfoStat. Centro de Transferencia InfoStat, FCA, Ing Agr. Felix Aldo Marrone 746–Ciudad Universitaria, Universidad Nacional de Córdoba, Córdoba, Argentina: 2018. Available online: http://www.infostat.com.ar (accessed on 6 May 2020).

48. Decaëns, T.; Jiménez, J.J. Earthworm communities under an agricultural intensification gradient in Colombia. *Plant Soil* **2002**, *2401*, 133–143. [CrossRef]

49. Domínguez, A.; Bedano, J.C.; Becker, A.R.; Arolfo, R.V. Organic farming fosters agroecosystem functioning in Argentinian temperate soils: Evidence from litter decomposition and soil fauna. *Appl. Soil Ecol.* **2014**, *83*, 170–176. [CrossRef]

50. Domínguez, A.; Bedano, J.C. The adoption of no-till instead of reduced tillage does not improve some soil quality parameters in Argentinean Pampas. *Appl. Soil Ecol.* **2016**, *98*, 166–176. [CrossRef]

51. Curry, J.P. Factors affecting the abundance of earthworms in soils. In *Earthworm Ecology*, 2nd ed.; Edwards, C.A., Ed.; CRC Press: Boca Raton, FL, USA, 2004; pp. 91–108, ISBN 0-8493-1819-X.

52. Fragoso, C.; Lavelle, P.; Blanchart, E.; Senapati, B.K.; Jimenez, J.J.; Martínez, M.A.; Decaens, T.; Tondoh, J. Earthworm communities of tropical agroecosystems: Origin, structure and influence of management practices. In *Earthworm Management in Tropical Agroecosystems*; Lavelle, P., Brussaard, L., Hendrix, P., Eds.; CABI: Wallingford Oxon, UK, 1999; pp. 27–55.

53. Felten, D.; Emmerling, C. Effects of bioenergy crop cultivation on earthworm communities—A comparative study of perennial (Miscanthus) and annual crops with consideration of graded land-use intensity. *Appl. Soil Ecol.* **2011**, *49*, 167–177. [CrossRef]

54. Díaz-Zorita, M.; Duarte, G.A.; Grove, J.H. A review of no-till systems and soil management for sustainable crop production in the subhumid and semiarid Pampas of Argentina. *Soil Till. Res.* **2002**, *65*, 1–18. [CrossRef]

55. Schmidt, O.; Clements, R.O.; Donaldson, G. Why do cereal–legume intercrops support large earthworm populations? *Appl. Soil Ecol.* **2003**, *22*, 181–190. [CrossRef]

56. Kautz, T.; Stumm, C.; Kösters, R.; Köpke, U. Effects of perennial fodder crops on soil structure in agricultural headlands. *J. Plant Nutr. Soil Sci.* **2010**, *173*, 490–501. [CrossRef]

57. Van Eekeren, N.; van Liere, D.; de Vries, F.; Rutgers, M.; de Goede, R.; Brussaard, L. A mixture of grass and clover combines the positive effects of both plant species on selected soil biota. *Appl. Soil Ecol.* **2009**, *42*, 254–263. [CrossRef]

58. Shipitalo, M.J.; Protz, R.; Tomlin, A.D. Effect of diet on the feeding and casting activity of Lumbricus terrestris and L. rubellus in laboratory culture. *Soil Biol. Biochem.* **1988**, *20*, 233–237. [CrossRef]

59. Molina, Y.; Mora, A.; Ramos, M.; Parra, L. Evaluación de dos especies leguminosas como abono verde. Cuenca alta del Río Chama, Mérida, Venezuela. *Rev. Forest. Venez.* **2011**, *55*, 183–193.

60. Vanzolini, J.I.; Galantini, J.; Agamennoni, R. Cultivos de cobertura de *Vicia villosa* Roth. en el valle bonaerense del Río Colorado. In *Contribuciones de los Cultivos de Cobertura a la Sostenibilidad de los Sistemas de Producción*; Álvarez, C., Quiroga, A., Santos, D., Bodrero, M., Eds.; INTA: La Pampa, Argentina, 2013; pp. 21–28, ISBN 978-987-679-177-9.

61. Forján, H.; Manso, L. Los Cereales de Invierno en la Secuencia de Cultivos. Su Aporte a la Sustentabilidad del Sistema de Producción. Available online: http://rian.inta.gov.ar/Boletines/Articulos/Documentos/Cereales_de_invierno_en_lasecuencia_de_cultivos.pdf (accessed on 6 May 2020).

62. Abail, Z.; Whalen, J.K. Corn residue inputs influence earthworm population dynamics in a no-till corn-soybean rotation. *Appl. Soil Ecol.* **2018**, *127*, 120–128. [CrossRef]

63. Darmawan, A.; Atmowidi, T.; Manalu, W.; Suryobroto, B. Land-use change on Mount Gede, Indonesia, reduced native earthworm populations and diversity. *Aust. J. Zool.* **2018**, *65*, 217–225. [CrossRef]

64. Domínguez, A.; Jiménez, J.J.; Ortíz, C.E.; Bedano, J.C. Soil macrofauna diversity as a key element for building sustainable agriculture in Argentine Pampas. *Acta Oecol.* **2018**, *92*, 102–116. [CrossRef]

65. Fragoso, C. Diversidad y patrones biogeográficos de las lombrices de tierra de México (Oligochaeta, Annelida). In *Minhocas na América Latina: Biodiversidade e Ecología*; Brown, G.G., Fragoso, C., Eds.; Embrapa Soja: Londrina, Brazil, 2007; pp. 107–124, ISBN 978-85-7033-019-2.

66. Eijsackers, H. Earthworms as colonizers of natural and cultivated soil environments. *Appl. Soil Ecol.* **2011**, *50*, 1–13. [CrossRef]

67. Frazão, J.; de Goede, R.G.; Brussaard, L.; Faber, J.H.; Groot, J.C.; Pulleman, M.M. Earthworm communities in arable fields and restored field margins, as related to management practices and surrounding landscape diversity. *Agric. Ecosyst. Environ.* **2017**, *248*, 1–8. [CrossRef]

68. Falco, L.B.; Momo, F.R.; Mischis, C.C. Ecología y biogeografía de las lombrices de tierra en la Argentina. In *Minhocas na América Latina: Biodiversidade e Ecología*; Brown, G.G., Fragoso, C., Eds.; Embrapa Soja: Londrina, Brazil, 2007; pp. 247–253, ISBN 978-85-7033-019-2.

69. Fragoso, C.; Brown, G.G.; Patron, J.C.; Blanchart, E.; Lavelle, P.; Pashanasi, B.; Senapati, B.; Kumar, T. Agricultural intensification, soil biodiversity and agroecosystem function in the tropics: The role of earthworms. *Appl. Soil Ecol.* **1997**, *6*, 17–35. [CrossRef]

70. Schmidt, O.; Curry, J.P.; Hackett, R.A.; Purvis, G.; Clements, R.O. Earthworm communities in conventional wheat monocropping and low-input wheat-clover intercropping systems. *Ann. Appl. Biol.* **2001**, *138*, 377–388. [CrossRef]

71. Fragoso, C.; Barois, I.; Gonzalez, C.; Arteaga, C.; Patron, J.C. Relationship between earthworms and soil organic matter levels in natural and managed ecosystems in the Mexican tropics. In *Soil Organic Matter Dynamics and Sustainability of Tropical Agriculture*, 1st ed.; Mulongoy, K., Merckx, R., Eds.; Wiley-Sayce Co-Publication: Chichester, UK, 1993; pp. 231–239, ISBN-13 978-0471939153.

Article

Comparison of the Effect of Perennial Energy Crops and Arable Crops on Earthworm Populations

Beata Feledyn-Szewczyk *, Paweł Radzikowski, Jarosław Stalenga and Mariusz Matyka

Department of Systems and Economics of Crop Production, Institute of Soil Science and Plant Cultivation—State Research Institute, Czartoryskich 8 Str., 24-100 Puławy, Poland; pradzikowski@iung.pulawy.pl (P.R.); stalenga@iung.pulawy.pl (J.S.); mmatyka@iung.pulawy.pl (M.M.)
* Correspondence: bszewczyk@iung.pulawy.pl; Tel.: +48-081-478-6803

Received: 1 October 2019; Accepted: 22 October 2019; Published: 24 October 2019

Abstract: The purpose of the study was to compare earthworm communities under winter wheat in different crop production systems on arable land—organic (ORG), integrated (INT), conventional (CON), monoculture (MON)—and under perennial crops cultivated for energy purposes—willow (WIL), Virginia mallow (VIR), and miscanthus (MIS). Earthworm abundance, biomass, and species composition were assessed each spring and autumn in the years 2014–2016 using the method of soil blocks. The mean species number of earthworms was ordered in the following way: ORG > VIR > WIL > CON > INT > MIS > MON. Mean abundance of earthworms decreased in the following order: ORG > WIL > CON > VIR > INT > MIS > MON. There were significantly more species under winter wheat cultivated organically than under the integrated system ($p = 0.045$), miscanthus ($p = 0.039$), and wheat monoculture ($p = 0.002$). Earthworm abundance was significantly higher in the organic system compared to wheat monoculture ($p = 0.001$) and to miscanthus ($p = 0.008$). Among the tested energy crops, Virginia mallow created the best habitat for species richness and biomass due to the high amount of crop residues suitable for earthworms and was similar to the organic system. Differences in the composition of earthworm species in the soil under the compared agricultural systems were proven. Energy crops, except miscanthus, have been found to increase earthworm diversity, as they are good crops for landscape diversification.

Keywords: soil biota; invertebrates; farming systems; bioenergy; biodiversity; wheat; ecosystem

1. Introduction

Earthworms constitute the largest component of animal biomass in the soil, and they are termed "soil engineers" [1] or even "ecosystem engineers" [2]. The diversity and abundance of earthworms are an important criterion of soil fertility [3]. Earthworms play a major role in the processes of decomposition of plant and animal organic residues, soil humus formation, creation of a crumbly soil structure, water regulation, pathogen and pest control, degradation of pollutants, as well as nitrogen binding [2,4]. Earthworms provide many ecosystem services, such as maintenance of proper soil structure, soil humus formation, nutrient cycling, erosion control, and biological crop protection [5].

Activity of earthworms and formation of burrows improve the aeration and water infiltration and therefore reduce surface runoff [6]. There are many mineral and organic ingredients in earthworm excrements that are beneficial for the growth and development of plants. These fractions are well mixed in worm casts, and the nutrients are present in a readily available form [1,5]. Soil passing through the digestive system of earthworms is enriched with beneficial microorganisms binding free nitrogen and activating phosphorus, which become available to plants. Both excreta and secretions (metabolic water and mucus) of earthworms contain plant growth stimulators (auxin, gibberellin, and cytokine), affecting the quality and quantity of the crop yield [7]. By pulling fallen leaves into the soil, foliar pathogens and pests are biologically degraded. Earthworms distribute insect-killing nematodes (*Steinernema* sp.) and fungi (*Beauveria bassiana*) in the soil, thus contributing to better natural regulation of soil-borne pests. Earthworms ingest organic residues of different C:N ratios, convert them to a lower C:N ratio, and finally contribute to carbon sequestration [5]. A rich earthworm fauna is a key in maintaining and safeguarding soil health and in fostering many essential ecosystem functions of soil.

Agricultural practices, such as tillage, crop rotation, cultivation of catch crops (defined here as additional fast-growing crops grown between successive plantings of main crops), the use of mineral fertilizers, and pesticides all have significant impacts on wild flora and fauna, including soil organisms [8–10]. The sensitivity of earthworms to unfavourable changes in agrocenosis makes them a good indicator of the ecological footprint of agricultural systems [11]. According to Pfiffner [5], the following measures are prerequisites for the flourishing of earthworms in agricultural soils: provision of sufficient food, abstinence from the use of pesticides harmful to earthworms, application of soil-conservation methods such as reduced tillage and no-till, avoidance of soil compaction and promotion of well-structured and aerated soils, appropriate fertilization, and balanced humus management within the crop rotation. Different farming systems can support or reduce biodiversity and soil settlement by earthworms. High-input conventional agriculture reduces the biodiversity of earthworms, whereas conservation and organic agriculture benefit this group of organisms [12–14].

Cultivation of perennial crops for energy purposes is still a new agricultural system so little scientific evidence appears to be available on the effects of these types of crops on the environment, including on earthworm populations [15,16]. The positive effect of energy crops on biodiversity is connected with the lower agrochemical input as compared to the intensive production used in annual crops [15,17–19]. Due to the unknown impact of many plant species used for energy purposes on the environment and biodiversity, there should be wide-ranging and long-term ecological monitoring conducted for these crops [19,20]. According to Verdade et al. [21], biodiversity monitoring programs are needed to help the decision-making process concerning the conflict between the expansion of energy crops and the conservation of biodiversity. These programs should take into account comparisons with neighbouring agricultural crops [22,23]. Such a comparison has been done in this current study.

Our hypothesis was that the cultivation of certain species of perennial energy crops stimulate earthworm diversity and abundance more than farming systems on arable land.

On this basis, the aim of the present research was to compare the impact of various agricultural systems on arable land (organic, integrated, conventional, and winter wheat monoculture) and the impact of cultivation of perennial energy crops (miscanthus, Virginia mallow, and willow) on earthworm diversity, abundance, and biomass under the conditions pertaining in Eastern Poland (Central Europe).

2. Materials and Methods

2.1. Site Description and Experimental Design

The assessment of earthworms was carried out as part of a long-term field experiment (1994–until now) with four crop production systems: organic (ORG), integrated (INT), conventional (CON), and monoculture-conventional (MON) (Table 1). Experimental plantations of perennial energy crops (2008–until now)—miscanthus (*Miscanthus* sp.) (MIS), Virginia mallow (*Sida hermaphrodita*) (VIR), and willow (*Salix viminalis*) (WIL)—have been established 70–100 m away from the crop production systems on the arable land described above. The experimental sites are located at the Agricultural Research Station of the Institute of Soil Science and Plant Cultivation—State Research Institute (IUNG-PIB) in Osiny (Poland, Lublin voivodeship, N: 51°28′, E: 22°30′) on *Haplic Luvisol* soil [24] with a texture of loamy sand. The chemical properties of the soil in the different farming systems are presented in Table 2. Average annual total precipitation of the experimental site was 586 mm with a mean air temperature of 7.5 °C (data for the years 1950–2013).

Table 1. Crop management in winter wheat in different farming systems and three perennial energy crops (2014–2016).

Items	Crop Production Systems on Arable Land				Energy Crops
	Organic (ORG)	Integrated (INT)	Conventional (CON)	Monoculture (MON)	
Crops	potato spring wheat + undersown crop clovers and grasses (1st year) clovers and grasses (2nd year) winter wheat + catch crop (mustard)	potato spring wheat + catch crop faba bean winter wheat + catch crop (mustard)	winter rape winter wheat spring wheat	winter wheat	miscanthus (MIS), Virginia mallow (VIR), willow (WIL)
Soil tillage	mouldboard ploughing				0
Organic fertilization	compost (30 t·ha^{-1}) under potato + catch crop	compost (30 t·ha^{-1}) under potato + 2 × catch crop	rape straw, winter wheat straw	wheat straw (every 2 years)	0
Mineral fertilization (kg ha^{-1}):	natural P and K fertilizers:				
N	0	85	140		80
P$_2$O$_5$	42	55	60		60
K$_2$O	75	75	80		80
Retardants	0	1–2 x	2 x		0
Fungicides	0	2 x	2–3 x		0
Weed control	weeder harrow 2–3 x	weeder harrow 1 x herbicides 1–2 x	herbicides 2–3 x		0

Table 2. Soil chemical properties of the 0–30 cm layer of *Luvisol*.

Cropping System	pH$_{KCl}$	C$_{org}$ (g kg^{-1} of Soil)	P$_{Egner}$	K$_{Egner}$	Mg
			(mg kg^{-1} of Soil)		
ORG [1]	5.65	9.9	40.3	64.0	69.3
CON	5.90	8.1	84.8	164.0	50.1
INT	5.75	8.1	85.4	134.1	41.9
MON	5.08	7.7	52.3	111.7	46.5
MIS	4.00	5.7	97.2	57.5	30.3
VIR	4.60	5.8	82.4	136.1	51.0
WIL	4.20	6.6	77.2	104.3	62.0

[1] ORG—organic; INT—integrated; CON—conventional; MON—monoculture; MIS—miscanthus; VIR—Virginia mallow; WIL—willow.

The tested farming systems on arable land were characterised by different crop rotations and agricultural management (Table 1). In the organic system, no synthetic pesticides and natural phosphorus (P) and potassium (K) fertilizers such as crude potassium salt or kainite as well as compost, applied once in a crop rotation, under potato (30 t ha^{-1}) were applied. High-input conventional systems included two variants: (1) 3-field crop rotation: winter oilseed rape, winter wheat, and spring wheat and (2) winter wheat monoculture. In both objects, crops were cultivated intensively, i.e., with high rates of synthetic mineral fertilizers and pesticides (Table 1). In the integrated system balanced mineral and organic fertilization (about 20–30% lower than in conventional system), adaptation to the crop requirements and soil fertility were used. Soil tillage was similar in all crop production systems; in general, it was a traditional plough system. Before sowing of winter crops, a four-furrow reversible plough was used at depths of 15–20 cm. Before sowing of spring crops, winter ploughing was carried out at a depth of minimum 20 cm. After late crops such as maize or potato or after cultivation of a catch crop, winter ploughing was done usually in the second half of November. For other crops harvested in July/August, winter ploughing was performed at the end of September. Before sowing, a compact seedbed cultivator was usually used. Number and time of ploughing treatments in particular systems depended on the crop rotation structure.

Crop protection consisted of nonchemical (mechanical) measures, and only a limited number of herbicides and other plant protection products were applied based on harmfulness thresholds. Most herbicides used were applied on crops, and only a few based on glyphosate were sprayed on stubble and incorporated into soil.

The analyses were carried out in the soil under winter wheat cv. Jantarka, which grew in all of these crop production systems. The area of each field was 1 ha so it was possible to manage them using real agricultural practices. In energy crops, no chemical crop protection measures have been performed during the study period. The area of the energy crop fields was from 200 m^2 (miscanthus and Virginia mallow) to 500 m^2 (willow).

2.2. Earthworm Collection

Earthworms were collected in the years 2014–2016, according to the method of soil blocks [25]. Samples were taken twice in each season: in spring (April) and in autumn (October), when soil temperature and moisture are usually suitable for earthworm activity. Soil blocks 25 cm × 25 cm × 25 cm (0.0625 m^2 of arable soil layer) were dug out of each field in 5 replications. The blocks were separated by at least 5 m from each other and 5 m away from the field margin to avoid the edge effect. Soil blocks were placed on a sheet on which the earthworms were caught and kept in collecting containers. The earthworms were transported in cool boxes to the laboratory, where they were washed, weighed, and preserved in 4% formalin for further investigations. In later terms, earthworms were classified into species according to an identification guide [26]. The species was recognized on the basis of morphological features. The structure of the mouth, distribution of bristles, number of segments, structure and location of the reproductive organs, and the location of secretory holes were thoroughly analyzed. Due to longer storage in the formalin solution, the colour of individuals was not taken into account. Both mature and young forms were considered, but only adult individuals were defined to the species [5]. Earthworm species richness, their abundance, and biomass (fresh weight) were taken into account as biodiversity indicators. Earthworm density and biomass were calculated per 1 m^2. For each investigated agricultural system, the results from all terms of collection were presented (3 years × 2 terms × 5 replications = 30 samples for each studied field).

2.3. Statistical Analyses

Because this trial was without replication, we chose to consider each sample point as a replicate, though this approach could have induced some bias because of pseudoreplications. We are, however, quite confident that, for several reasons, our sampling design allowed us to use this approach: (i) sample points within a field were far enough away from each other (20 m) to ensure replicate independence, and (ii) the possibility that the sample fields were affected by confounding factors due to limited randomization cannot be excluded but was limited as the trial was evenly affected by the same management before the trial setup, topographic and pedologic gradients were controlled, and a preliminary assessment of the trial spatial heterogeneity was found to be very low within the block soil heterogeneity. This methodological approach has been used earlier by Henneron et al. [14]. There are examples of field trials that are not replicated, for example, one of the oldest long-term experiments established in 1843 Rothamsted (UK) [27], but, with the help of specific statistical methods, permits reliable comparisons. Lack of replications is due to different reasons, but usually, it is caused by physical constraints such as land availability or plot size. Sometimes financial or social limitations are also important [28].

In order to check the normality of the distributions, the Shapiro–Wilk test was used. The obtained data sets were characterized by a non-normal distribution; therefore, the nonparametric Kruskal–Wallis rank test was used to assess the significance of differences in the examined features at the significance level $p \leq 0.05$. Statistical analyses were performed using Statistica 10 software (Stat. Soft. Inc., Tulsa, OK, USA).

In order to group the earthworm communities, a cumulative hierarchical classification was done using MVSP 3.1. software, Kovach Computing Services, Anglesey, Wales [29]. The quantitative Sorensen's similarity index (Percent Similarity) was used to classify similarities between earthworms under different farming systems and types of energy crops [30].

In order to classify the samples based on earthworm species composition and species based on their participation in the samples, ordination techniques were used [31]. As first, Detrended Correspondence Analysis (DCA) was applied, which is recommended for the preliminary ordering of data [31,32]. The length of variance gradient calculated in this analysis characterizes the data structure and constitutes the criterion for selecting further ordination methods to assess the significance of the tested environmental or agrotechnical factors. Due to the fact that the length of the first axis gradient in DCA analysis was lower than 2 standard deviations, which showed that the distribution of species was not compatible with the Gaussian curve, linear method Principal Component Analysis (PCA) was used to perform direct ordination [33]. The results of this ordination were presented graphically on diagram (PCA diplot). The analyses were performed in the Canoco 4.5 program [32].

3. Results

3.1. Species Richness and Abundance

A total of 11 species of earthworms were recorded in the compared agricultural systems (Table 3). The largest number of species occurred in the willow (10 species), and the lowest number was in the high-input crop production systems: wheat monoculture and the conventional system (7 species). In the organic system, there were many juvenile unspecified individuals (Lumbricidae sp.). It should be noted that *Allobophora chlorotica* was found only in crop production systems on arable land and that *Proctodrilus antipai* was only found under perennial energy plants.

Table 3. Earthworm species and abundance of individuals (indv. m^{-2}) in soil under winter wheat cultivated in different crop production systems and in perennial energy crops (2014–2016).

No	Species	Crop Production Systems on Arable Land				Perennial Energy Crops		
		ORG [1]	INT	CON	MON	MIS	VIR	WIL
1.	*Aporrectodea caliginosa*	13.9	21.9	16.5	21.3	9.6	10.1	9.1
2.	*Allolobophora chlorotica*	0	0.5	0	0	0	0	0
3.	*Aporrectodea georgii*	0	3.2	1.1	0	0	0	0.5
4.	*Aporrectodea longa*	1.6	0	0	0.5	2.7	0.5	1.1
5.	*Aporrectodea rosea*	5.3	4.8	2.7	2.1	8.5	3.7	9.6
6.	*Proctodrilus antipai*	0	0	0	0	1.6	5.9	4.8
7.	*Lumbricus rubellus*	1.1	0.5	0	0	0	2.7	1.1
8.	*Lumbricus terrestris*	15.5	2.1	7.5	1.6	2.7	13.3	10.7
9.	Lumbricidae sp.	24.0	15.5	15.5	3.2	6.9	16.5	33.6
10.	*Octolasion cyaneum*	4.3	2.1	10.7	1.1	3.2	1.6	0.5
11.	*Octolasion lacteum*	22.9	5.9	10.7	4.8	7.5	9.1	3.2
	Total species number	8	9	7	7	8	9	10
	Abundance (indv. m^{-2}) (mean ± SE)	88.6 ± 10 b [2]	56.5 ± 9 ab	64.7 ± 10 ab	34.6 ± 6 a	42.7 ± 7 a	63.4 ± 9 ab	74.2 ± 18 b

[1] ORG—organic; INT—integrated; CON—conventional; MON—monoculture; MIS—miscanthus; VIR—Virginia mallow; WIL—willow. [2] Different letters indicate significant differences between treatments according to the Kruskal–Wallis test at $p \leq 0.05$ ($n = 30$).

In crop production systems on arable land, the largest number of earthworm individuals (88.6 indv. m^{-2}) was recorded in the soil under winter wheat cultivated in organic system (Table 3). Over twice less individuals (35 indv. m^{-2}) were found in monoculture of winter wheat. Earthworms density decreased in the order of ORG > CON > INT > MON. Among compared energy crops, willow had the largest earthworm abundance (74 indv. m^{-2}, only 16% less than in the organic system) while the smallest was found in miscanthus field (43 indv. m^{-2}, 23% more than in wheat monoculture).

Total earthworm abundance was significantly higher in the organic system as compared to winter wheat monoculture ($p = 0.001$; $z = 3.85$) and to miscanthus ($p = 0.008$; $z = 3.65$) (Table 3). There were no significant differences between the organic system and willow in earthworm density.

Earthworm species number per sample was higher under winter wheat cultivated organically than under monoculture ($p = 0.002$; $z = 2.68$), than the integrated system ($p = 0.045$; $z = 3.64$), and than under miscanthus ($p = 0.039$; $z = 3.24$) (Figure 1). Earthworm diversity in the organic system and in Virginia mallow did not differ significantly. Among the tested energy crops, differences in the species richness were found between miscanthus and Virginia mallow ($p = 0.032$; $z = 2.55$).

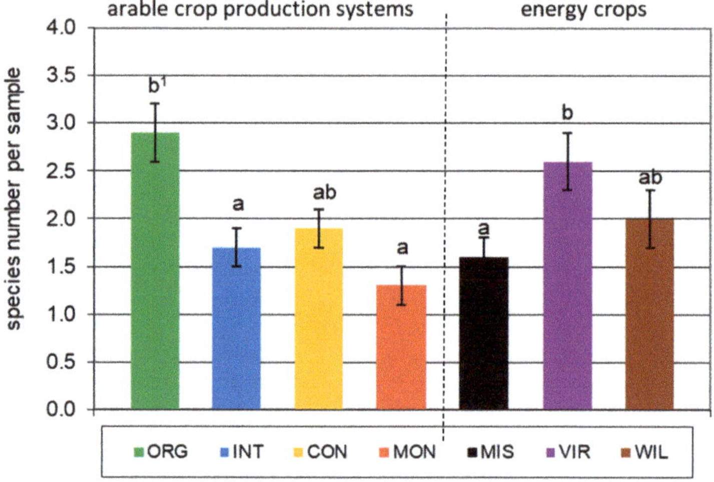

Figure 1. Average earthworm species number per sample (0.0625 m^2) in winter wheat cultivated in four crop production systems on arable land and perennial energy crops (average for 2014–2016, mean ± SE, n = 30). ORG—organic; INT—integrated; CON—conventional; MON—monoculture; MIS—miscanthus; VIR—Virginia mallow; WIL—willow. [1] Different letters indicate significant differences between treatments according to the Kruskal–Wallis test at $p \leq 0.05$.

The biomass of earthworms was 3 times larger in the organic system in comparison with the wheat monoculture, miscanthus, and willow systems (Figure 2). There were no significant differences between the organic, conventional, and integrated systems, which could be the result of using compost and catch crops twice in the integrated system and of using straw ploughing in the conventional system (Table 1).

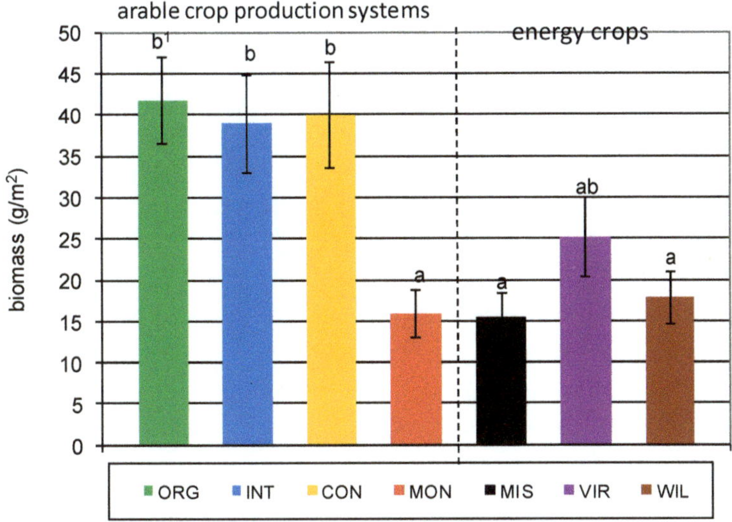

Figure 2. Earthworm biomass (fresh weight g m^{-2}) in winter wheat cultivated in four crop production systems on arable land and perennial energy crops (average for 2014–2016, mean ± SE, n = 30). ORG—organic; INT—integrated; CON—conventional; MON—monoculture; MIS—miscanthus; VIR—Virginia mallow; WIL—willow. [1] Different letters indicate significant differences between treatments within arable systems and energy crops according to the Kruskal–Wallis test at $p \leq 0.05$.

3.2. The Relationships between Agricultural Systems and Species Composition

In the organic system, the indeterminate individuals (Lumbricidae sp.), mainly juvenile (27%) and *O. lacteum* (26%), constituted the most numerous groups (Figure 3). The highest density of *L. terrestris* was found in the organic system and Virginia mallow. The earthworm communities in the integrated system and the wheat monoculture system were dominated by *A. caliginosa* (39% and 60% share, respectively). The highest share of *A. rosea* were observed in earthworm communities under miscanthus and willow. In the soil under willow and Virginia mallow, earthworms unspecified for species, mainly epigeic juveniles of low biomass, dominated. In the miscanthus cultivation, *A. caliginosa* and *A. rosea* were the most numerous.

In order to confirm the relationship between land use and the abundance of earthworm species, the ordination method PCA (Principal Component Analysis) was used (Figure 4).

Along the gradients of the axes of particular crop production systems, species most closely associated with a given type of farming and energy crop were grouped together (Figure 4). *O. lacteum* was most closely connected with crops cultivated organically. *L. terrestris*, *A. longa*, *L. rubellus*, and unidentified Lumbricidae sp. were located close to the gradients of the axes of the organic system and Virginia mallow, which indicates that they were characteristic for both crop production systems. *P. antipai* and *A. rosea* were strongly positively correlated with willow. Species on the left-hand side of the diagram in Figure 4—*A. caliginosa*, *O. cyaneum*, *A. chlorotica*, and *A. georgii*—were associated with more intensive farming systems (CON, MON, and INT) and miscanthus.

Gradients of the systems and the species distribution in Figure 4 showed similarity of the earthworm communities for the organic system and Virginia mallow (right, upper site of the diagram) and dissimilarities from other, more intensive crop production systems (INT and MON) as well as miscanthus (left, down site of the diagram). Along the gradient of axis I, the highest positive correlation between the tested systems and the location of species occurred for willo and negative occurred for wheat monoculture. The most positively correlated with the gradient of axis 2 was the organic system while that negatively correlated was willow (Table A1, Figure 4).

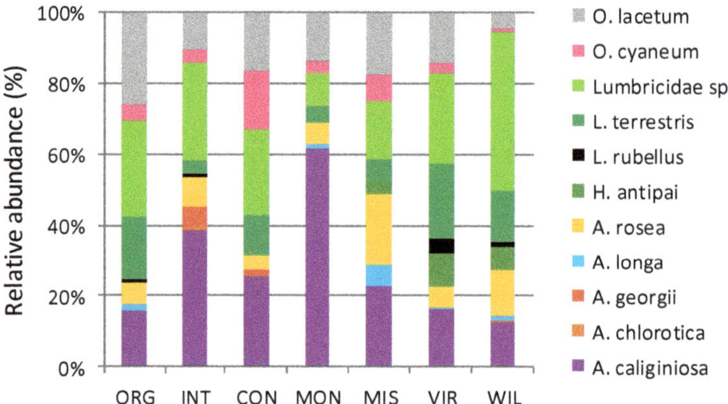

Figure 3. Relative abundance of earthworms species in communities in different agricultural production systems. ORG—organic; INT—integrated; CON—conventional; MON—monoculture; MIS—miscanthus; VIR—Virginia mallow; WIL—willow.

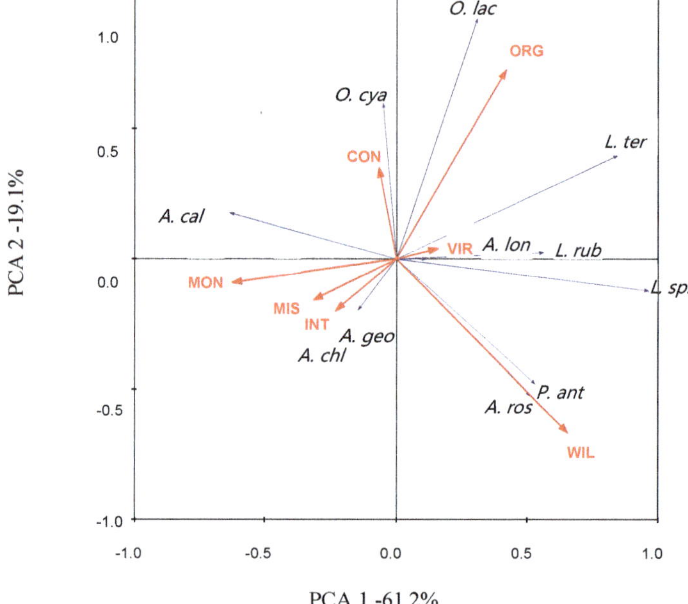

Figure 4. Ordination diagram of agricultural systems and earthworm species abundance in relation to the first and second axes of PCA (PCA biplot). ORG—organic; INT—integrated; CON—conventional; MON—monoculture; MIS—miscanthus; VIR—Virginia mallow; WIL—willow.

The hierarchical classification confirmed the similarity of the earthworm communities in the organic and Virginia mallow systems and their dissimilarity from integrated system, wheat monoculture, and miscanthus (Figure 5). In these analyses, the dissimilarity of earthworm communities under miscanthus from the other two energy crops and their similarity to wheat monoculture and the integrated system have been shown.

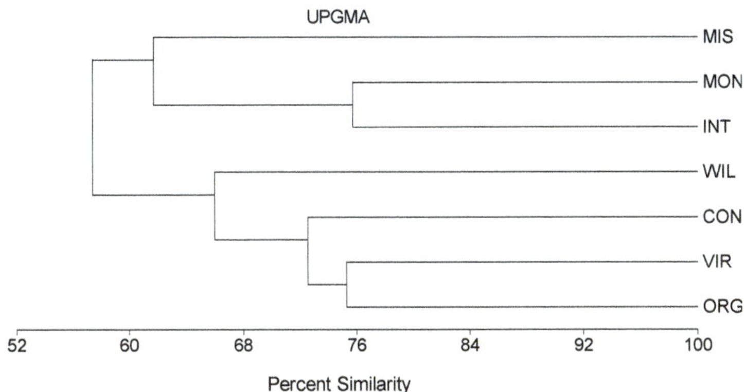

Figure 5. The results of the hierarchical cumulative classification using the unweighted pair group method with arithmetic mean (UPGMA) of samples representing earthworm abundance in different types of land use using the quantitative Sorensen's coefficient of similarity. ORG—organic; INT—integrated; CON—conventional; MON—monoculture; MIS—miscanthus; VIR—Virginia mallow; WIL—willow.

4. Discussion

4.1. Species Richness and Abundance of Earthworms

In tested farming systems, on arable land, and among energy crops, 11 species of earthworms were recorded. Neirynck et al. [34] and Pfiffner [5] have shown that, generally, in cropland, only from 1 to 11 species are commonly found from among the about 40 earthworm species occurring in Central Europe. Most of them are species with a wide range and large adaptation abilities. The most common species in Poland are *Dendrobaena octaedra* (Sav.), *Lumbricus terrestris* L., *L. rubellus* Hoffm., *Aporrectodea caliginosa* (Sav.), *Aporrectodea rosea* (Sav.), and *Eiseniella tetraedra* (Sav.) [26].

The relative abundance of earthworms depends upon soil type, topography, and vegetation, and it is influenced by land use [1,34]. According to Lavelle [35], in terrestrial ecosystems, density of earthworms may reach 10^6 ha $^{-1}$ and their biomass may reach 2 t ha $^{-1}$. They are present everywhere except in arid and frozen regions. In Central Europe, about 120–140 earthworms per 1 m^2 of arable soil are considered satisfactory in terms of soil fertility [5]. In our own research, a smaller number of earthworms were recorded, i.e., between 88 indv. m^{-2} in the organic system up to 35 indv. m^{-2} in monoculture. The lower earthworm abundance could be the result of being sandier, having higher acidity, and having low humus content of the investigated soils compared to the other European countries, which are not favourable to earthworms [5]. Very low earthworm abundance on sandy soils was also observed on conventional arable dairy farm situated in Peer in the sands of the Campine region (Belgium) [36].

Almost all of the earthworm species found are typical of the region and land use [26]. *A. caliginosa* was most abundant in arable systems. It is a species found in a wide range of environments. It is highly tolerant of habitat quality, including acidification and low levels of organic matter. The species tolerates intensive tillage to some extent, making it the most common species on arable land. Other highly represented species, *L. terrestris*, *O. lacteum*, *O. cyaneum*, *A. longa*, *A. rosea*, and *L. rubellus*, have slightly higher habitat requirements and are typical rather for meadows and pastures habitats than for arable land [35]. The species with the highest tolerance for low pH is *A. chlorotica*, which was found to have one individual in a cereal monoculture. Among the rare species, *A. georgii* was found. This species was previously found in the neighbouring region of the Mazovian voivodeship [26]. A rare, South European species of the earthworm, *P. antipai*, is usually found on heavy soils in the Vistula river valley, whereas in the studies, it was found in poor light soil. Heavy hydrogenic soils also lie in the farm serving the experiments, which suggests that both species could be transferred with the soil on agricultural equipment.

4.2. The Effect of Crop Rotation and Catch Crops

The number of earthworms depends on the availability and quality of organic matter in the soil [6,26]. On the arable land, manure and compost as well as the protein-rich residues of legumes (*Fabaceae*) are particularly preferred by earthworms. A diversified crop rotation with long-lasting and deep-rooted catch crops rich in clover or green manure crops as well as diversified crop residues are essential to maintain or increase earthworm populations [5]. This was confirmed by our own results showing that earthworm species richness and density were significantly higher in the organic system than in other systems (by 126% higher mean number of species and by 156% more individuals than in wheat monoculture). The higher species number and abundance of earthworms in the organic system could be the effect of the diversified crop rotation with clover and grasses mixture as well as of the compost and catch crop used, which provided food for earthworms. In a study by Schmidt et al. [37], a wheat-clover cropping system supported earthworm communities that were twice as large and had between one and five times more earthworm species than that found under conventional wheat mono-cropping. The above authors recorded between one and five more earthworm species in the wheat-clover system than in the pure wheat system.

Increased earthworm density was also observed as a result of mulching of crop residues on the field surface, especially during winter [5,38]. Post-harvest residues are an important source of organic matter, and they also affect the microclimatic properties of the soil. In mulched soil, there are more earthworms feeding close to the soil surface. In addition, such soil is more resistant to freezing than the soil without plant cover, which has an impact on the mortality of earthworms during winter, late autumn, and spring [12]. In our study, larger number and biomass of earthworms in the conventional system than in the integrated system could have been caused by a large biomass of post-harvest residues from the incorporation of wheat and rape straw.

The content and type of humus in soil are some of the most important factors determining the species composition of earthworms. *L. terrestris* has a tendency to occur in soil rich in organic carbon [39]. In our own research, *L. terrestris* was the most frequent in wheat in the organic system and Virginia mallow, where the content of organic matter was the highest. According to Schmidt et al. [37], wheat-clover cropping especially favoured species belonging to the epigeic and epigeic/anecic ecological groups such as *L. rubellus* and juvenile *Lumbricus* and was also observed in our own study, whereas *A. caliginiosa* may occur even in soils poor in carbon [39]. In the presented study, *A. caliginosa* was the most numerous in the intensive systems on arable land: integrated, wheat monoculture, conventional, as well as willow. A higher share of species typical for meadows and pastures such as *L. terrestris*, *O. lacteum*, *O. cyaneum*, and *A. longa* in the organic system is the result of 1.5 years of clover and grass cultivation preceding winter wheat in rotation. During this period, species building permanent vertical tunnels are not disturbed by soil tillage and receive some fresh plant residues on the soil surface, which promote development of such earthworm communities [35].

4.3. The Influence of Different Farming Systems

The effect of farming systems on earthworm species richness is a result of different agricultural practices, such as of crop rotation, crop protection, and fertilization.

Application of pesticides in conventional crop production systems can decrease density and biomass of earthworm population [12]. Pesticides may disrupt enzymatic processes, increase individual mortality, decrease fecundity and growth, or even change individual behavior such as feeding rate [5]. Anectic earthworms such as *L. terrestris* are most susceptible to surface application of pesticides, which, in our research, corresponds with a higher density of this species in the organic system in comparison with other agricultural systems where pesticides were used. Since *L. terrestris* forms permanent burrows, it does not come into contact with subsurface soil in its burrows. However, this species collects plant residues and pulls it into its tunnels, which is why it is directly exposed to the use of pesticides. On the contrary, endogenic species such as *A. caliginosa*, which continuously extend their burrows as they feed in the subsurface soil, are the most susceptible when toxic pesticides are incorporated into the soil [5]. According to Irmler's research [12], the change from conventional to organic management has positively influenced species that form deep tunnels such as *L. terrestris*. The conversion from conventional to organic did not significantly affect the species penetrating shallow, horizontal tunnels, such as *A. caliginiosa* and *A. rosea*.

Herbicides probably do not damage earthworms directly, but they can reduce earthworm populations by decreasing availability of organic matter coming from weeds on the soil surface [40]. According to some authors, soil fauna is more threatened by the use of insecticides than herbicides. Active substances, such as carbofuran, forat, and terbufos, used to control soil-dwelling pests are also extremely toxic to earthworms [41].

Some synthetic mineral fertilizers, especially ammonium sulfate, can be harmful to earthworm populations, probably due to an acidifying affect [5]. On the contrary, the use of manure increases both the number and biomass of earthworms in arable land [42]. This was confirmed by our own study where the earthworm density was significantly the highest in the organic system. Similarly, Pfiffner's and Mader's [43] studies showed that, in conventional fields, where mineral fertilization and integrated plant protection were applied, the number and biomass of earthworms was about 40% lower compared to the fields where mineral-organic fertilization and integrated plant protection was applied. However, in the organic system, the number of earthworms was additionally 80% higher and the biomass was 40% higher in comparison to the object with mineral-organic fertilization and integrated plant protection. Moreover, the quality and quantity of manure could be important factors affecting the earthworm population due to salinity stress [44]. In our study, earthworm abundance under the organic system was higher than under the integrated, conventional, and wheat monoculture systems by 56%, 36%, and 156%, respectively. Similarly, Pfiffner and Luka [45] found about a 50% higher abundance of earthworms in an organic system as compared to integrated ones. Comparisons between organic and conventional systems have shown from an 80% [43] to a 400% higher earthworm density in the organic system [14]. In our research, the biomass of earthworms was 3 times larger in soil under the organic system than in the wheat monoculture, miscanthus, and willow. Similarly, Henneron et al. [14] found a 3 times larger biomass of earthworms in an organic system in comparison with high-input conventional system. In the study by Pfiffner and Mäder [43], the difference in earthworm biomass between organic and conventional systems with mineral fertilization and Integrated Pest Management (IPM) was about 75% and only 35% when the conventional system with mixed mineral and organic fertilization and IPM was compared. Stopping the use of synthetic fertilizers and pesticides are important factors that stimulate the population size and condition of earthworms in the organic system [5,14,26]. In more intensive systems, simplified crop rotation and high input of synthetic fertilizers and pesticides negatively affected the earthworm populations [12].

4.4. The Influence of Perennial Energy Crops

The influence of perennial energy crops on fauna diversity depends on cultivated species and agricultural practices [23,46]. Our study showed that mean earthworm species number, density, and biomass in energy crops were intermediate between wheat in organic system and high-input, intensive wheat monoculture and were dependent on the type of energy crop. Among tested energy crops, Virginia mallow created the best habitat in terms of species richness and biomass due to high amount of crop residues suitable for earthworms. The small amount of plant residues in miscanthus resulted in low earthworm indicators and high similarity to intensive crop production systems. In Felten and Emmerling's [16] research, the number of earthworm species under miscanthus was placed between intensively cultivated crops (rapeseed, cereals, and maize) and grassland/fallow. In comparison to annual cropping systems, miscanthus plantation enhanced higher densities and diversity of soil invertebrates but not of ground-dwelling invertebrates [47]. In miscanthus stand, earthworm diversity and abundance were improved in arable soils although biomass may be reduced through poorer food quality [48]. Miscanthus leaf litter does not provide a particularly useful food resource due to its low-nitrogen, high-carbon nature [49], and earthworms feeding on this kind of low-nitrogen material have been found in other studies to lose overall mass [50]. In contrast, though, the extensive litter cover at ground level under miscanthus compared to the bare soil under annual cereals was suggested to be a potentially significant advantage for earthworms in soil surface moisture retention and protection from predation [48].

Another study confirmed that miscanthus created a poorer habitat for fauna than did willow [51]. In our own research, earthworm density was the highest in the soil under willow. Studies conducted in Great Britain and Sweden confirmed the positive impact of willow on the diversity of invertebrates in comparison with crops cultivated in intensive conventional systems on arable land [19,52].

According to Hedde et al. [53], the change in land use from typical annual crops to perennial energy plants resulted in an increase in the density of invertebrates in the soil, which may be caused by a smaller amount of synthetic fertilizers and chemical plant protection chemicals used, with no significant changes in richness and species composition of tested invertebrates. In our study, earthworm indicators were dependent on the type of energy crop which correspond with our working hypothesis. In the cultivation of energy crops, more earthworms depended on the presence of mulch. *L. rubellus* as well as species that burrow deep tunnels, *L. terrestris*, *A. longa*, and *O. lacteum*, were present, which was also a consequence of the lack of tillage. The composition of earthworm species found in energy plants resembled that found in orchards [35].

Further research should be warranted to design and assess innovative cropping systems including the range of candidate bioenergy crops, possibly grown in alternative lands, and also in the face of future climate changes [54].

5. Conclusions

It can be concluded that the organic crop production system with a diversified crop rotation including grass-clover mixtures, catch crops, and manure application in the conditions of Eastern Poland (Central Europe) supported the largest diversity and abundance of earthworms in comparison to the high-input cereal monoculture and integrated systems. The organic system favoured the population of *L. terrestris* and juvenile Lumbricidae. The intensification of agricultural production by simplified crop rotation and the input of synthetic fertilizers and chemical plant protection products caused a decrease in the number of species and abundance of earthworms. *A. caliginiosa* dominated in the earthworm community in the monoculture of wheat. On plantations of energy crops, the earthworm population indices were located between the organic system and the high-input, intensive wheat monoculture. The effect of energy crop cultivation on earthworm abundance and ecosystem services which are provided depends on the respective crop species. Among the compared energy crops, Virginia mallow created the best habitat for earthworms. In miscanthus, earthworm community was the poorest and the most similar to wheat monoculture. The composition of earthworm species found in energy plants resembled that found in orchards. Proper management of energy crops can support biodiversity and ecosystem services supplied by earthworms, such as humus production..

Author Contributions: Conceptualization, M.M., J.S., and B.F.-S.; methodology, B.F.-S., P.R., and J.S.; validation, M.M. and J.S.; investigation, B.F.-S. and P.R.; data curation, B.F.-S. and P.R.; writing—original draft preparation, B.F.-S. and P.R.; writing—review and editing, M.M. and J.S.; visualization, B.F.-S. and P.R.; supervision, M.M. and J.S.; project administration, J.S. and M.M.; funding acquisition, J.S. and M.M.

Funding: The study was carried out under the statutory project of IUNG-PIB no 1.13. (2014–2017). This work has been cofinanced by the National (Polish) Centre for Research and Development (NCBiR) entitled "Environment, agriculture and forestry" project: BIOproducts from lignocellulosic biomass derived from MArginal land to fill the Gap in Current national bioeconomy, No. BIOSTRATEG3/344253/2/NCBR/2017.

Conflicts of Interest: The authors declare no conflict of interest. The funders had no role in the design of the study; in the collection, analyses, or interpretation of data; in the writing of the manuscript; or in the decision to publish the results.

Appendix A

Table A1. Inter set correlations of agricultural systems variables with Principal Component Analysis (PCA) axes in study of earthworms communities.

Variables	Axis I	Axis II	Axis III	Axis IV
ORG [1]	0.4171	0.7252	−0.0595	0.0732
INT	−0.2282	−0.1979	−0.6013	0.0376
CON	−0.0664	0.3486	−0.2498	−0.5203
MON	−0.6226	−0.0908	−0.1363	0.4262
MIS	−0.3108	−0.1575	0.6736	−0.5513
VIR	0.1573	0.0399	0.5006	0.6283
WIL	0.6536	−0.6674	−0.1273	−0.0937

[1] ORG—organic; INT—integrated; CON—conventional; MON—monoculture, MIS—miscanthus, VIR—virginia mallow, WIL—willow.

References

1. Kooch, Y.; Jalilvand, H. Earthworms as ecosystem engineers and the most important detritivors in forest soils. *Pak. J. Biol. Sci.* **2008**, *11*, 819–825. [CrossRef] [PubMed]

2. Blouin, M.; Hodson, M.E.; Delgado, E.A.; Baker, G.; Brussaard, L.; Butt, K.R.; Dai, J.; Dendooven, L.; Peres, G.; Tondoh, J.E.; et al. A review of earthworm impact on soil function and ecosystem services. *Eur. J. Soil Sci.* **2013**, *64*, 161–182. [CrossRef]

3. Van Groenigen, J.W.; Lubbers, I.M.; Vos, H.M.J.; Brown, G.G.; De Deyn, G.B.; van Groenigen, K.J. Earthworms increase plant production: A meta-analysis. *Sci. Rep.* **2014**, *4*, 6365. [CrossRef] [PubMed]

4. Kuntz, M.; Berner, A.; Gattinger, A.; Mäder, P.; Pfiffner, L. Influence of reduced tillage on earthworm and microbial communities under organic arable farming. *Pedobiologia* **2013**, *56*, 251–260. [CrossRef]

5. Pfiffner, L. Earthworms—Architects of fertile soils. Their significance and recommendations for their promotion in agriculture. In *Technical Guide on Earthworms*; International edition; Research Institute of Organic Agriculture: Frick, Switzerland, 2014; Available online: https://shop.fibl.org/chde/mwdownloads/download/link/id/624/ (accessed on 21 March 2019).

6. Lee, K.E. Earthworms and sustainable land use. In *Earthworm Ecology and Biogeography in North America*; Hendrix, P.F., Ed.; CRS Press: Boca Raton, FL, USA, 1995; pp. 215–234.

7. Sinha, R.K.; Valani, D.; Chauhan, K.; Agarwal, S. Embarking on a second green revolution for sustainable agriculture by vermiculture biotechnology using earthworms: Reviving the dreams of Sir Charles Darwin. *J. Agric. Biotechnol. Sust. Dev.* **2010**, *2*, 113–128.

8. Jordan, D.; Miles, R.J.; Hubbard, V.C.; Lorenz, T. Effect of management practices and cropping systems on earthworm abundance and microbial activity in Sanborn Field: A 115-year-old agricultural field. *Pedobiologia* **2004**, *48*, 99–110. [CrossRef]

9. Moos, J.H.; Schrader, S.; Paulsen, H.M.; Rahmann, G. Occasional reduced tillage in organic farming can promote earthworm performance and resource efficiency. *App. Soil Ecol.* **2016**, *103*, 22–30. [CrossRef]

10. Stalenga, J.; Brzezińska, K.; Stańska, M.; Błaszkowska, B.; Czekała, W.; Feledyn-Szewczyk, B.; Gutkowska, A.; Hajdamowicz, I.; Kaliszewski, G.; Kazuń, A.; et al. *Code of Good Agricultural Practices Supporting Biodiversity*; Monograph. II revised ed.; IUNG-PIB: Puławy, Poland, 2016; p. 250.

11. Fründ, H.C.; Graefe, U.; Tischer, S. Earthworms as bioindicators of soil quality. In *Biology of Earthworms*; Karaca, A., Ed.; Springer: Berlin/Heidelberg, Germany, 2011; pp. 261–278.

12. Irmler, U. Changes in earthworm populations during conversion from conventional to organic farming. *Agric. Ecosyst. Environ.* **2010**, *135*, 194–198. [CrossRef]

13. Flohre, A.; Rudnick, M.; Traser, G.; Tscharntke, T.; Eggers, T. Does soil biota benefit from organic farming in complex vs. simple landscape? *Agric. Ecosyst. Environ.* **2011**, *141*, 210–214. [CrossRef]

14. Henneron, L.; Bernard, L.; Hedde, M.; Pelosi, C.; Villenave, C.; Chenu, C.; Bertrand, M.; Girardin, C.; Blanchart, E. Fourteen years of evidence for positive effects of conservation agriculture and organic farming on soil life. *Agron. Sust. Dev.* **2014**, *35*, 169–181. [CrossRef]

15. Pedroli, B.; Elbersen, B.; Frederiksen, P.; Grandin, U.; Heikkila, R.; Krogh, P.H.; Izakovicova', Z.; Johansen, A.; Meiresonne, L.; Spijker, J. Is energy cropping in Europe compatible with biodiversity? Opportunities and threats to biodiversity from land-based production of biomass for bioenergy purposes. *Biomass Bioenergy* **2013**, *55*, 73–86. [CrossRef]

16. Felten, D.; Emmerling, C. Effects of bioenergy crop cultivation on earthworm communities—A comparative study of perennial (*Miscanthus*) and annual crops with consideration of graded land-use intensity. *Appl. Soil Ecol.* **2011**, *49*, 167–177. [CrossRef]

17. Sage, R.B. Short rotation coppice for energy: Towards ecological guidelines. *Biomass Bioenergy* **1998**, *15*, 39–47. [CrossRef]

18. Börjesson, P. Environmental effects of energy crop cultivation in Sweden—I: Identification and quantification. *Biomass Bioenergy* **1999**, *16*, 137–154. [CrossRef]

19. Cunningham, M.D.; Bishop, J.D.; McKay, H.V.; Sage, R.B. *ARBRE Monitoring—Ecology of Short Rotation Coppice*; URN 04/961, DTI; Game Conservancy Trust: Fording Bridge, UK, 2004; 157p.

20. European Environmental Agency. *How much Bioenergy can Europe Produce without Harming the Environment*; EEA Report 7/2006; European Environmental Agency: København, Denmark, 2006; 67p.

21. Verdade, L.M.; Piña, C.I.; Rosalino, L.M. Biofuels and biodiversity: Challenges and opportunities. *Environ. Dev.* **2015**, *15*, 64–78. [CrossRef]

22. Kovacs-Lang, E.; Simpson, I.C. *Biodiversity Measurements and Indicators for Long-Term Integrated Monitoring*; No LIMITS, Report No 6; Institute of Ecology and Botany of the Hungarian Academy of Sciences, Vácrátót & Centre for Ecology and Hydrology: Grange-over-Sands, UK, 2000; 24p.

23. Britt, C. *Methodologies for Ecological Monitoring in Bioenergy Crops. A Review and Recommendations*; Defra Project NF0408; ADAS Contract Report for the Department for Environment, Food and Rural Affairs: London, UK, 2003; 63p.

24. IUSS Working Group WRB. *World Reference Base for Soil Resources*, 2nd ed.; World Soil Resources Reports, No. 103; FAO: Rome, Italy, 2006; 132p.

25. Zhang, S.; Chao, Y.; Zhang, C.; Cheng, J.; Li, J.; Ma, N. Earthworms enhanced winter oilseed rape (*Brassica napus* L.) growth and nitrogen uptake. *Agric. Ecosyst. Environ.* **2010**, *139*, 463–468. [CrossRef]

26. Kasprzak, K. *Soil Oligochaetes, III. Family: Earthworms (Lumbricidae)*; Keys for marking the invertebrates of Poland; PAN, Institute of Zoology: Warsaw, Poland, 1986; Volume 6, 186p. (In Polish)

27. Leigh, R.A.; Johnston, A.E. (Eds.) *Long-Term Experiments in Agricultural and Ecological Sciences*; CAB Int.: Wallingford, UK, 1994; 448p.

28. Machado, S.; Petrie, S.E. Symposium—Analysis of unreplicated experiments: Introduction. *Crop Sci.* **2006**, *46*, 2474–2475. [CrossRef]

29. Kovach, W.L. *MVSP Version 3*; Kovach Computing Services: Anglesey, Wales, 2011; 112p.

30. Magurran, A.E. *Ecological Diversity and its Measurement*; Princeton University Press: Princeton, NJ, USA, 1988; 179p.

31. Lepš, J.; Šmilauer, P. *Multivariate Analysis of Ecological Data Using CANOCO*; Cambridge University Press: Cambridge, UK, 2003; 269p.

32. Ter Braak, C.J.F.; Smilauer, P. *CANOCO Reference Manual and CanoDraw for Windows User's guide: Software for Canonical Community Ordination (version 4.5)*; Microcomputer Power: Ithaca, NY, USA, 2002; 55p.

33. Van der Maarel, E. Multivariate analysis in plant ecology. In *Numerical Methods in Studying the Structure and Functioning of Vegetation*; Kaźmierczak, E., Ed.; V School and XLVI Geobotanical Seminar of the Polish Botanical Society; University Nicholas Copernicus in Torun: Torun, Poland, 1998; pp. 65–108.

34. Neirynck, J.S.; Mirtcheva, G.; Lust, N. Impact of *Tilia patyphllos* Scop., *Fraxinus excelsior* L., *Acer pseudoplatanus* L., *Quercus robur* L. and *Fagus sylvatica* L. on earthworm biomass and physico-chemical properties of loamy topsoil. *Forest Ecol. Man.* **2000**, *133*, 275–286. [CrossRef]

35. Lavelle, P. The structure of earthworm communities. In *Earthworm Ecology*; Satchell, J.E., Ed.; Springer: Dordrecht, The Netherlands, 1983; pp. 449–466. [CrossRef]

36. Valckx, J.; Hermy, M.; Muys, B. Indirect gradient analysis at different spatial scales of prorated and non-prorated earthworm abundance and biomass data in temperate agro-ecosystems. *Eur. J. Soil Biol.* **2006**, *42*, 341–347. [CrossRef]

37. Schmidt, O.; Curry, J.P.; Hackett, R.A.; Purvis, G.; Clements, R.O. Earthworm communities in conventional wheat monocropping and low-input wheat clover intercropping systems. *Ann. Appl. Biol.* **2001**, *138*, 377–388. [CrossRef]

38. Schmidt, O.; Curry, J.P. Population dynamics of earthworms (Lumbricidae) and their role in nitrogen turnover in wheat and wheat clover cropping systems. *Pedobiologia* **2001**, *45*, 174–187. [CrossRef]

39. Poier, K.R.; Richter, J. Spatial distribution of earthworms and soil properties in an arable loess soil. *Soil Biol. Biochem.* **1992**, *24*, 1601–1608. [CrossRef]

40. Paluszek, J. Criteria for Assessing the Physical Quality of Poland's Arable Soils. In *Acta Agrophysica*; Institute of Agrophysics Polish Academy of Sciences: Lublin, Poland, 2011; 138p. (In Polish)

41. McLaughlin, A.; Mineau, P. The impact of agricultural practices on biodiversity. *Agric. Ecosyst. Environ.* **1995**, *55*, 201–212. [CrossRef]

42. Doran, J.W.; Werner, M.R. Management and soil biology. In *Sustainable Agriculture in Temperate Zones*; Francis, C.A., Flora, C.B., King, L.D., Eds.; Wiley: New York, NY, USA, 1990; pp. 205–230.

43. Pfiffner, L.; Mäder, P. Effects of biodynamic, organic and conventional systems on earthworm populations. *Biol. Agric. Hort.* **1997**, *15*, 3–10. [CrossRef]

44. Hurisso, T.T.; Davies, J.G.; Brummer, J.E.; Stromberger, M.E.; Stonaker, F.H.; Kondratieff, B.C.; Booher, M.R.; Goldhamer, D.A. Earthworm abundance and species composition in organic forage production systems of northern Colorado receiving different soil amendments. *Appl. Soil Ecol.* **2011**, *48*, 219–226. [CrossRef]

45. Pfiffner, L.; Luka, H. Earthworm populations in two low-input cereal farming systems. *Appl. Soil Ecol.* **2007**, *37*, 184–191. [CrossRef]

46. Perttu, K.L. Ecological, biological balances and conservation. *Biomass Bioenergy* **1995**, *9*, 107–116. [CrossRef]

47. Hedde, M.; van Oort, F.; Boudon, E.; Abonnel, F.; Lamy, I. Responses of soil macroinvertebrate communities to Miscanthus cropping in different trace metal contaminated soils. *Biomass Bioenergy* **2013**, *55*, 122–129. [CrossRef]

48. McCalmont, J.P.; Hastings, A.; McNarama, N.P.; Richter, G.M.; Robson, P.; Donnison, I.S.; Clifton-Brown, J. Environmental costs and benefits of growing Miscanthus for bioenergy in the UK. *GCB Bioenergy* **2015**. [CrossRef]

49. Heaton, E.A.; Dohleman, F.G.; Long, S.P. Seasonal nitrogen dynamics of Miscanthus x giganteus and Panicum virgatum. *GCB Bioenergy* **2009**, *1*, 297–307. [CrossRef]

50. Abbott, I.; Parker, C. Interactions between earthworms and their soil environment. *Soil Biol. Biochem.* **1981**, *13*, 191–197. [CrossRef]

51. Lewandowski, I.; Clifton-Brown, J.C.; Scurlock, J.M.O.; Huisman, W. Miscanthus: European experience with a novel energy crop. *Biomass Bioenergy* **2000**, *19*, 209–227. [CrossRef]

52. Fry, D.A.; Slater, F.M.; Reboud, X. The effect on plant communities and associated taxa of planting short rotation willow coppice in Wales. *Asp. Appl. Biol.* **2008**, *90*, 287–293.

53. Hedde, M.; van Oort, F.; Renouf, E.; Thénard, J.; Lamy, I. Dynamics of soil fauna after plantation of perennial energy crops on polluted soils. *Appl. Soil Ecol.* **2013**, *66*, 29–39. [CrossRef]

54. Gabrielle, B.; Bamière, L.; Caldes, N.; De Cara, S.; Decocq, G.; Ferchaud, F.; Loyce, C.; Pelzer, E.; Perez, Y.; Wohlfahrt, J.; et al. Paving the way for sustainable bioenergy in Europe: Technological options and research avenues for large-scale biomass feedstock supply. *Renew. Sust. Energ. Rev.* **2014**, *33*, 11–25. [CrossRef]

 *agronomy*

Communication

Do Long-Term Continuous Cropping and Pesticides Affect Earthworm Communities?

Kinga Treder *, Magdalena Jastrzębska, Marta Katarzyna Kostrzewska and Przemysław Makowski

Department of Agroecosystems, Faculty of Environmental Management and Agriculture, University of Warmia and Mazury in Olsztyn, Plac Łódzki 3, 10-718 Olsztyn, Poland; magdalena.jastrzebska@uwm.edu.pl (M.J.); marta.kostrzewska@uwm.edu.pl (M.K.K.); przemyslaw.makowski@uwm.edu.pl (P.M.)
* Correspondence: kinga.treder@uwm.edu.pl

Received: 4 March 2020; Accepted: 16 April 2020; Published: 20 April 2020

Abstract: Earthworm species composition, the density of individuals, and their biomass were investigated in spring barley and faba bean fields in a long-term (52-year) experiment conducted at the Production and Experimental Station in Bałcyny, in north-eastern Poland (53°40′ N; 19°50′ E). Additionally, post-harvest residues biomass, soil organic matter (SOM), and soil pH were recorded. The above traits were investigated using two experimental factors: I. cropping system—continuous cropping (CC) vs. crop rotation (CR) and II. pesticide plant protection: herbicide + fungicide (HF+) vs. no plant protection (HF−). A total of three species of *Lumbricidae* were found: *Aporrectodea caliginosa* (Sav.) in both crops, *Aporrectodea rosea* (Sav.) in spring barley, and *Lumbricus terrestris* (L.) in faba bean. The density and biomass of earthworms were unaffected by experimental treatments in spring barley fields, whereas in faba bean CC increased and HF+ decreased earthworm density and biomass in comparison with CR and HF− respectively. Total post-harvest residues in faba bean fields were higher under CC in relation to CR and under HF+ compared with HF− treatment in both crops. Compared to CR, CC increased soil pH in spring barley fields and decreased in faba bean fields. Experimental factors did not affect SOM. Earthworm density and biomass were positively correlated with SOM content.

Keywords: soil organic matter; soil pH; post-harvest residues; crop rotation; *Hordeum vulgare* L.; *Vicia faba* L. ssp. *minor*

---

1. Introduction

Earthworms are strategic invertebrates in agroecosystems. The drillosphere, composed of horizontal and vertical burrows and casts created by earthworms, significantly affects soil structure and enhances gas exchange, water infiltration, and root penetration across the soil profile [1–4]. Earthworms improve the content of soil organic matter, contribute to humus formation processes and form a mull soil by burrowing large quantities of surface organic matter to belowground and relocating soil from depths to the top by casting. These invertebrates have an impact on the structure, concentration, and activity of soil microbial communities involved in organic matter decomposition and mineralization [5,6]. Earthworms casts are characterized by higher pH, C, Ca^{2+}, Mg^{2+}, and K^+ contents than surrounding soil aggregates and incorporate nutrients available for plants [7,8]. The N mineral availability increases with earthworm abundance [9,10]. Earthworms, through their interaction with microorganisms, are essential factors influencing soil organic carbon and its dynamics [10,11]. In the presence of earthworms, greater production of plant growth regulators was observed [12,13]. The non-negligible role of earthworms on improving plant tolerance to parasitic nematodes and a reduction in the severity of take-all disease was also reported [14–16]. The above-mentioned benefits of earthworm activity contribute to plant growth and production as seen by aboveground biomass and crop yield increases [17–19].

Density, diversity, structure, and activity of earthworm populations in agroecosystems are dependent on agricultural management [20]. Intensification of agricultural practices based on multiple tillage treatments, simple crop sequence, minor organic fertilization, and chemical methods of plant protection have a negative effect on earthworm populations. Long-term, intensive, and deep tillage can decline earthworm (mainly anecic) abundance [21–24] whereas shallow plowing with residue mixing and conservation cultivation techniques can increase their number [25–27]. Organic fertilizers such as manure, crop residues, and mulch are the source of food supply for soil biota and have a positive impact on their populations [28,29]. The conclusion from the literature on the effect of pesticides on earthworm populations is still ambiguous. The effect of pesticides on soil organisms is closely related to their active substances and doses. Some of them, especially fungicides and insecticides, are toxic or lethal to earthworms and cause a decrease in cocoon production and density of juveniles, a delay in growth and a mortality increase [3,30–32]. Herbicides were also found to have an adverse impact on earthworms by causing histological changes in their body tissues and increasing mortality [33–37]. However, earthworms can develop adaptation mechanisms against the toxic effect of pesticides [38,39].

The importance of earthworms in agroecosystems is well-recognized, but the long-term effects of continuous cropping and pesticide use on earthworm populations are much less documented. The objective of this study was to examine the impact of the above-mentioned factors on species composition, density, and biomass of earthworms, post-harvest residue biomass, soil organic matter, and soil pH. The alternative hypothesis assumed that long-term continuous cropping and chemical plant protection have an impact on earthworm communities was tested against the null hypothesis that the above factors do not affect the analyzed parameters.

2. Materials and Methods

2.1. Experimental Design and Crop Management

The field experiment was initiated in autumn 1967 in the Production and Experimental Station in Bałcyny in north-eastern Poland (53°40′ N; 19°50′ E). During the first five years, nine crops were sown in a continuous cropping system (growing of the same crop on the same field each year). In 1972, two crop rotations were included to analyze continuous cropping impact. Crop rotation varied throughout the experiment. Currently, twelve crops in continuous cropping and in two crop rotations (growing different crops one after the other on the same field) are being sown. The crop rotations are A. sugar beet, maize, spring barley, peas, winter rape, and winter wheat; B. potato, oats, fiber flax, winter rye, faba bean, and winter triticale.

Fertilization in the first sixteen years was the only mineral. Since 1983 farmyard manure was included in doses: 30 t ha^{-1} on potato/sugar beet field and 15 t ha^{-1} every three years in a continuous cropping system. Mineral fertilizers are applied in terms and doses respective to each crop's needs.

In one part of every crop field, no pesticides have been ever applied, which provides a unique chance to study no plant protection effect after 52-years of a continuous cropping and crop rotation system. To have a comparison for these results on the other parts of each crop field, herbicides (since 1972 till now) and fungicides (since 1983 till now) have been included. Throughout the experiment, the use of pesticides has been updated according to The Institute of Plant Protection National Research Institute recommendations.

The results presented in this paper were based on two crops differing in their biology and agricultural importance: spring barley (cultivar Radek) and faba bean (cultivar Amigo). Spring barley is grass with a short root system and short vegetation period whereas faba bean has a deep, well-developed root system with nitrogen-fixing nodules. Faba bean, by nitrogen fixation and high mass of residues, increases soil biological activity, organic matter content, porosity and soil moisture, which has an impact on earthworm communities [40].

Basic agricultural data for spring barley and faba bean in 2019 are presented in Table 1.

Table 1. Basic agricultural data for spring barley and faba bean in 2019.

Item	Spring Barley	Faba Bean
Soil tillage system	Plow tillage (with crop residues incorporation after harvest)	
Mineral fertilisation		
-P_2O_5 (kg/ha)	70 [1]	60 [1]
-K_2O (kg/ha)	100 [1]	100 [1]
-N (kg/ha)	70 (50 [1] + 20 [2])	40 [1]
Plant protection herbicides	Mustang 306 SE; florasulam + 2,4-D EHE; 0.5 l/ha; stem elongation *	Corum 502.4 SL; bentazon + imazamox; 1.25 l/ha; 16 leaf unfolded Dash HC; fatty acid esters + alkoxylated alcohols-phosphate esters; 0.6 l/ha; 16 leaf unfolded
fungicides	Capalo 337.5 SE; fenpropimorph + epoxiconazole + metrafenone; 1.5 l/ha; stem elongation Amistar 250 SC; azoxystrobin; 0.6 l/ha; flowering Artea 330 EC; propiconazole + cyproconazole; 0.4 l/ha; flowering	Dithane NeoTec 75 WG; mancozeb; 2.0 kg/ha; flower buds visible outside leaves

[1]—before sowing, [2]—at stem elongation stage; * trade name; active ingredient; rate; crop growth stage.

Two experimental factors were investigated, each with two levels: I. cropping system: continuous cropping (CC) vs. crop rotation (CR), II. plant protection: herbicide + fungicide (HF+) vs. no plant protection (HF−). In both spring barley and faba bean, particular experimental treatments (CC-HF+, CC-HF−, CR-HF+, CR-HF−) were performed in 3 replications (i.e., plots). From each plot 3 samples were taken. That brought the total to 9 samples for each treatment. Each plot size was 27 m^2.

2.2. Soil Characteristics

The experiment was established on Luvisol medium soil, derived from light loam lying on loamy sand. At the beginning of the lasting rotation (2016) soil contained an average (mg kg^{-1}) of available forms of phosphorous—289.3, potassium—258.5, magnesium—55.0, total nitrogen—800 with 1.1% C$_{org}$, and pH—5.7.

2.3. Meteorological Data

The climate in this region is temperate humid, with annual total precipitation around 587.5 mm and a mean annual air temperature of 7.9 °C (data for the years 1981–2015). Weather conditions of the July–September 2019 period are presented in Figure 1. High air temperature in combination with small rainfall before sampling in August could have reduced earthworm activity and biomass. In September, the rainfall and temperature were more favorable for earthworms [41].

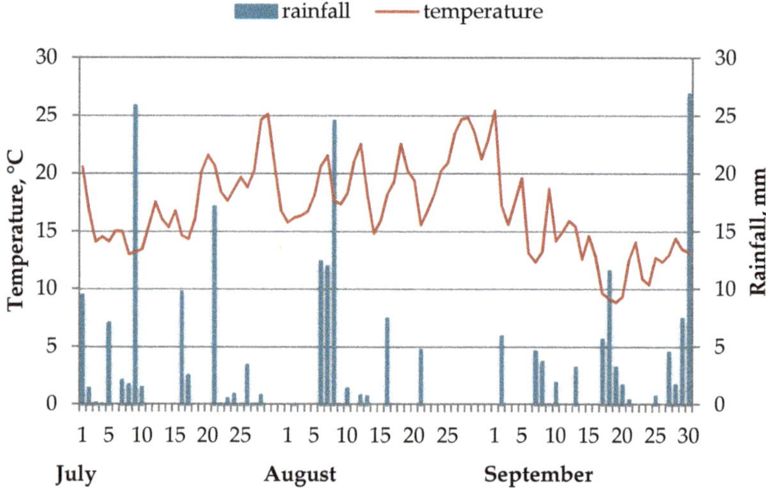

Figure 1. Daily temperature and rainfall before sampling.

2.4. Sampling

2.4.1. Post-Harvest Residue Sampling and Preparation

Just after harvest, the post-harvest residues (crop and weed bottom stalks with roots) were removed from a surface area of 0.40 m^2 and a depth of 0.30 m in three replications from each plot. Residues were transported to the laboratory in plastic bags. Crop and weed residues were separated, washed under tap water and air-dried for several days. Crop residues were split into shoots and roots. The dry mass of post-harvest residues was weighed.

2.4.2. Soil Sampling and Preparation

Three soil samples from each plot were selected on the days of earthworm collection. Soil samples were taken in three replications from each plot from a 0–30 cm depth with a hand-held twisting probe (Egner's soil sampler) and returned in plastic bags to the laboratory. Stones and plant residues were removed from the soil. Soil samples were air-dried in plastic trays and then passed through a 2 mm sieve to prepare homogenous samples.

Soil Organic Matter Analysis

The soil organic matter (SOM) was determined using the loss-on-ignition method [42]. Soil samples (each of mass 10 g) were oven-dried in crucibles at 105 °C for 2 h to remove hygroscopic water and then cooled and weighed. Afterward, soil samples were heated at 360 °C for 2 h and cooled and weighed. SOM (%) was calculated as the mass lost during combustion:

$$\text{SOM} = [(\text{soil weight 105 °C} - \text{soil weight 360 °C})/\text{soil weight 105 °C}] \times 100 \qquad (1)$$

Soil pH Determination

Soil pH was determined using the potentiometric method after water extraction (pH (H$_2$O)). In glass beakers, 10 g of the air-dried and sieved soil samples were mixed with 25 mL of distilled water and shaken. After two hours of equilibration, electrical conductance was measured in soil-water suspension for 10 min using a Hanna HI 221 (Hanna Instruments) pH meter.

2.4.3. Earthworm Sampling and Identification

On 12 plots of each crop, three soil samples (25 cm × 25 cm × 40 cm depth) were sampled after each crop harvest: spring barley—1.08.2019 and faba bean—26.09.2019. Earthworms were selected from soil blocks by the hand-sorting method. In the laboratory, earthworms were rinsed, dried, weighed (data refers to live biomass), narcotized in 35% ethyl alcohol and preserved in 4% formalin and 75% ethyl alcohol. Clitelated individuals were classified into species level, and juveniles were classified into species or genera level by external morphology using keys [41].

2.5. Statistical Analysis

The data were analyzed statistically by a two-factorial analysis of variance ANOVA in the STATISTICA (data analysis software system), version 12, StatSoft. Homogeneous groups were estimated by Duncan's test at a $p < 0.05$. The Shapiro–Wilk W-test was used for testing the normality of variable distribution and Levene's test for homogeneity of variance. The correlation coefficients were calculated to measure the strength and direction of the relationship between the variables. The coefficients were determined based on the data from all treatments, separately for spring barley and faba bean.

3. Results and Discussion

3.1. Post-Harvest Residues

In spring barley fields, the cropping system did not affect the dry mass of post-harvest residue (Table 2). However, CC related to CR increased faba bean mass of shoots, weeds and total residues. Positive influence of CC on faba bean shoots residues biomass was an unexpected result because growing faba bean in the same field year after year leads to the accumulation of autotoxic compounds (mainly phenolic acids) in the soil that inhibits plants growth [43]. In previous studies based on this experiment, Rychcik and Zawiślak [44] reported lower faba density under CC than CR treatment. Though, the lower density of plants may result in their higher biomass. This effect was reported by Kotecki in the experiment with faba bean, were biomass of one plant increased with density decrease [45]. Post-harvest biomass of weeds is the aftereffect of weeds density. The current consensus is that crop rotation, in contrast to continuous cropping, decreases weeds density, which was confirmed by Rychcik in faba bean fields [46].

HF+ compared to HF− treatment resulted in higher production of residues in both crops: in spring barley—shoots, in faba bean—shoots and roots. Furthermore, the biomass of weeds residues was significantly higher in HF− cereal and legume fields in relation to HF+. The presented results are obvious, considering the effects of herbicide use, which was in line with previous studies [46].

In spring barley fields, CC-HF+ increased the total residue biomass in comparison to CC-HF−, while there was no difference between CR-HF+ and CR-HF−. CR-HF− resulted in less spring barley shoots mass production related to CC-HF+ but it was higher than under CC-HF−. The biomass of weed residues under CC-HF− and CR-HF− was higher than in CR-HF+. The achieved results are in agreement with a previous study concerning weed infestation of spring barley in this experiment [47].

Similar results were noted in faba bean fields. The total residue biomass in CC-HF− and CR-HF+ was significantly lower than in CC-HF+ and higher in relation to CR-HF− interaction.

Moreover, HF+ in relation to HF− increased the biomass of faba bean shoots residues in CR, but not as strong as with CC. In relation to other treatments, CC-HF−increased the mass of weed residues. The biomass of faba bean roots was unaffected by treatment interactions.

Table 2. Dry mass of post-harvest residues (t ha^{-1}).

Treatments	Total	Shoots	Roots	Weeds
Spring Barley				
cropping system				
CR *	2.82a ** ± 0.15	2.07a ± 0.10	0.62a ± 0.07	0.13a ± 0.04
CC	2.74a ± 0.39	2.04a ± 0.33	0.49a ± 0.09	0.20a ± 0.03
plant protection				
HF−	2.37b ± 0.21	1.70b ± 0.16	0.45a ± 0.07	0.23a ± 0.02
HF+	3.18a ± 0.26	2.42a ± 0.21	0.66a ± 0.07	0.11b ± 0.04
interaction				
CR-HF−	2.74ab ± 0.27	2.02b ± 0.14	0.52a ± 0.10	0.21a ± 0.03
CR-HF+	2.89ab ± 0.17	2.12ab ± 0.18	0.71a ± 0.05	0.05b ± 0.01
CC-HF−	2.00b ± 0.16	1.37c ± 0.05	0.38a ± 0.11	0.24a ± 0.02
CC-HF+	3.48a ± 0.46	2.71a ± 0.28	0.61a ± 0.14	0.16ab ± 0.05
Faba Bean				
cropping system				
CR	1.24b ± 0.15	0.42b ± 0.09	0.47a ± 0.13	0.35b ± 0.08
CC	1.93a ± 0.18	0.70a ± 0.17	0.49a ± 0.14	0.73a ± 0.13
plant protection				
HF−	1.29b ± 0.17	0.29b ± 0.05	0.26b ± 0.07	0.74a ± 0.13
HF+	1.87a ± 0.19	0.82a ± 0.11	0.70a ± 0.10	0.34b ± 0.07
interaction				
CR-HF−	0.97c ± 0.05	0.25c ± 0.07	0.26a ± 0.09	0.46b ± 0.12
CR-HF+	1.51b ± 0.18	0.59b ± 0.05	0.69a ± 0.17	0.23b ± 0.06
CC-HF−	1.62b ± 0.21	0.34c ± 0.06	0.27a ± 0.13	1.01a ± 0.04
CC-HF+	2.23a ± 0.16	1.06a ± 0.09	0.72a ± 0.16	0.45b ± 0.12

* CR—crop rotation, CC—continuous cropping, HF−—no plant protection, HF+—herbicide + fungicide; ** values with different letters vary significantly (Duncan's test, $p < 0.05$), x ± sem—mean ± standard error of mean.

3.2. Soil Organic Matter Content

Experimental factors and their interactions did not affect the SOM in spring barley and faba bean fields (Table 3). These results are in line with the previous study based on this experiment where the C_{org} content in spring barley was comparable to faba bean fields in both cropping systems [48]. The same amounts of manure during every six-year-lasting rotation were applied in all treatments, so this may be the major cause of undifferentiated SOM levels. In contrast to the presented results, some authors [49–51] assert that crop rotation, especially with legumes, gives preferential conditions for soil C increase.

Table 3. Soil organic matter (SOM) (%).

Treatments	Spring Barley	Faba Bean
cropping system		
CR *	1.86 ** ± 0.10	1.98 ± 0.04
CC	1.88 ± 0.06	2.08 ± 0.08
plant protection		
HF−	1.89 ± 0.11	2.11 ± 0.05
HF+	1.85 ± 0.03	1.95 ± 0.07
interaction		
CR-HF−	1.84 ± 0.21	2.03 ± 0.06
CR-HF+	1.87 ± 0.08	1.93 ± 0.06
CC-HF−	1.93 ± 0.11	2.20 ± 0.03
CC-HF+	1.83 ± 0.03	1.96 ± 0.13

* CR—crop rotation, CC—continuous cropping, HF−—no plant protection, HF+—herbicide + fungicide; ** values do not differ significantly (Duncan's test, $p < 0.05$), x ± sem—mean ± standard error of mean.

3.3. Soil pH

In comparison with CR, CC increased soil pH in spring barley (Table 4). Hickman [52] reported lower pH in maize continuous cropping than in maize—wheat and maize—soybean rotations and soybean continuous cropping. The author suggested that these results may be explained by the long-term use of anhydrous ammonia in maize crops fields. HF+ lowered soil pH in relation to HF−. Spring barley under HF+ achieved higher yields than HF− (data not published). With higher yields, larger amounts of macroelements were removed from the soil. In spring barley fields, the interaction between cropping system and plant protection was proved. Under CR-HF+ treatment lower soil pH than under CR-HF− was noted, while there was no difference in soil pH under CC-HF− and CC-HF+. In faba bean fields, the values of soil pH were lower in CC than in CR. Dinitrogen-fixing legumes, including faba bean, are considered to generate soil acidification by releasing H+ to rhizosphere [53–55]. Lee [56] reported that continuous legume cultivation increased soil acidity. Comparably, Williams [57] noted soil pH decrease in a long-term experiment with clover pastures.

Table 4. Soil pH.

Treatments	Spring Barley	Faba Bean
	cropping system	
CR *	6.37b ** ± 0.11	6.73a ± 0.06
CC	6.89a ± 0.04	6.32b ± 0.03
	plant protection	
HF−	6.74a ± 0.09	6.55a ± 0.09
HF+	6.51b ± 0.17	6.50a ± 0.12
	interaction	
CR-HF−	6.56b ± 0.06	6.75a ± 0.05
CR-HF+	6.18c ± 0.16	6.72a ± 0.13
CC-HF−	6.92a ± 0.06	6.35b ± 0.03
CC-HF+	6.85ab ± 0.05	6.28b ± 0.05

* CR—crop rotation, CC—continuous cropping, HF−—no plant protection, HF+—herbicide + fungicide; ** values with different letters vary significantly (Duncan's test, $p < 0.05$), x ± sem—mean ± standard error of mean.

HF+ caused pH decrease only in spring barley fields. Spring barley under HF+ produced higher yields than under HF− (own data not published). Thus with higher yields, larger amounts of macroelements were removed from the soil.

In faba bean fields, no interaction of cropping system and plant protection on soil pH was revealed.

3.4. Earthworm Species Richness and Structure

In the experimental fields, only three species of *Lumbricidae* were found: *Aporrectodea caliginosa* (Sav.), *Aporrectodea rosea* (Sav.), and *Lumbricus terrestris* (L.) (Figure 2). The same species composition was reported by Valchovski [58] in cultivated and non-cultivated Vertic Luvisol. The species richness of earthworms in agroecosystems is usually lower than in natural ecosystems and depends mainly on soil type, soil humidity, organic matter content, cultivation treatments, and crop type [59–61]. In both crops, the most numerous species was *Aporrectodea caliginosa* (Sav.), which occurs commonly in arable lands in all temperate zones [62–64]. This endogenic earthworm can adapt to unfavorable environmental conditions like low soil moisture, low organic matter content, or tillage practices [22,65–67]. Although *Aporrectodea rosea* (Sav.) is very common in agroecosystems in different crops as well as in pastures [20,62,68], in the current study it was recorded only in spring barley fields under CR-HF+ and CC-HF− treatments. It is worth noting that *Lumbricus terrestris* (L.) was found only in CR-HF+ and CC-HF+ faba bean fields. Edwards [62] suggests that one of the limiting factors for *Lumbricus terrestris* (L.) abundance is soil organic matter. Faba bean plants have a well-developed, extensive fibrous root system, which gives preferential treatment to anecic earthworms.

In both crops, the majority of the earthworms found were juveniles representing the genera *Aporrectodea*.

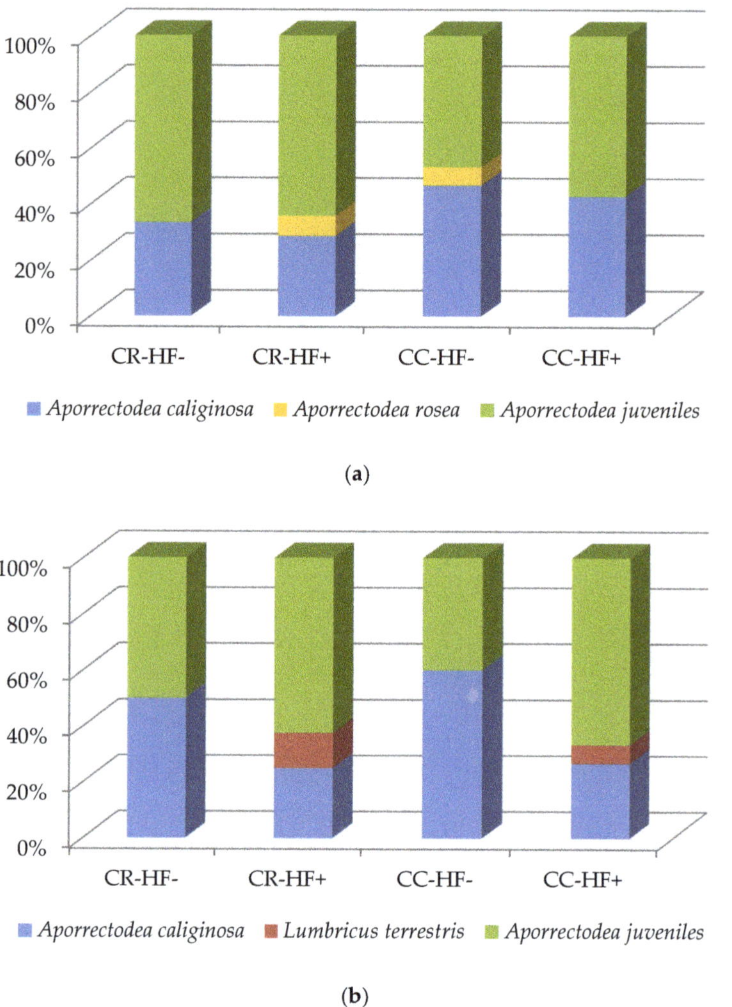

(a)

(b)

Figure 2. Relative abundance (%) of earthworm species (based on individuals m^{-2}) on (**a**) spring barley and (**b**) faba bean (B) fields; CR—crop rotation, CC—continuous cropping, HF-—no plant protection, HF+—herbicide + fungicide.

3.5. Earthworm Density and Biomass

The density of individuals and biomass of collected earthworms was lower than reported in other works [64,69,70], but comparable with results achieved in other researchers conducted in The Experimental Station in Bałcyny [71–74]. Experimental treatments did not affect the density or biomass of earthworms in spring barley fields (Table 5). Spring crops do not create favorable environmental conditions for earthworm abundance because of low organic matter input and agrotechnical works on fields in early spring when the activity of earthworm communities increases.

In faba bean fields, both experimental treatments affected earthworm abundance. In comparison with CR, CC increased the density of individuals and biomass of earthworms. The reason for this could be the higher mass of post-harvest residues. In a study by Edwards [62], the abundance of earthworms was two times higher in fields with continuous wheat (for 136 years) than with wheat—root crops. The negative influence of HF+ on the earthworm population in relation to HF− on faba bean fields was apparent in the decrease in individual density and their biomass. Many authors reported a reduction in growth, biomass loss, decreased cocoon production or higher mortality after pesticide application [75–79]. Biomass reduction may be an effect of reduced food intake as a strategy to avoid contamination. Nonetheless, the response to agrochemicals differs depending on earthworm species, the concentration of toxic substances, environmental conditions (soil type, temperature, humidity, organic matter content, etc.) and the duration of exposure [77,80,81]. CC-HF− treatment had a positive influence on earthworm density and biomass in relation to other experimental treatments.

Table 5. Density (individuals m^{-2}) and biomass (g m^{-2}) of earthworms in spring barley and faba bean fields.

Treatments	Spring Barley		Faba Bean	
	Individuals	Biomass	Individuals	Biomass
	cropping system			
CR *	15.3 ± 1.90	7.4 ± 0.94	8.0b ** ± 2.07	4.3b ± 1.28
CC	14.7 ± 3.37	5.9 ± 2.14	13.3a ± 3.21	9.0a ± 2.07
	plant protection			
HF−	14.0 ± 2.47	5.9 ± 1.52	15.3 a ±2.81	9.4a ± 2.19
HF+	16.0 ± 2.91	7.5 ± 1.77	6.0b ± 0.89	4.0b ± 0.67
	interaction			
CR-HF−	18.7 ± 1.33	8.9 ± 1.04	10.7b ± 3.52	5.5b ± 2.42
CR-HF+	12.0 ± 6.92	6.0 ± 1.08	5.3b ± 1.33	3.0b ± 0.81
CC-HF−	9.3 ± 2.67	2.8 ± 1.06	20.0a ± 2.30	13.2a ± 1.83
CC-HF+	20.0 ± 4.61	9.1 ± 3.46	6.7b ± 1.33	4.9b ± 0.85

* CR—crop rotation, CC—continuous cropping, HF-—no plant protection, HF+—herbicide + fungicide; ** values with different letters vary significantly (Duncan's test, $p < 0.05$), no letters—no significant differences, x ± sem—mean ± standard error of mean.

3.6. Relationship Between Earthworm Abundance and Post-Harvest Residues, SOM, and Soil pH

In spring barley fields, no relationship between earthworm abundance and post-harvest residues, SOM or soil pH was observed (Table 6). A strong positive association of SOM and earthworm abundance was noted in faba bean fields This suggests that organic matter left in the field by a legume, characterized by high protein content may be beneficial for earthworm populations. The above is in the line with Kladivko [82], where a higher density of earthworms in continuous soybean than in continuous corn was reported. The association between soil organic matter content and the presence of the earthworms was also observed in other findings [83,84]. Soil pH and dry mass of post-harvest residues were not significantly correlated with earthworm density or biomass.

Table 6. The correlation coefficient (r) between density and biomass of earthworms with SOM, pH, and post-harvest residues of spring barley and faba bean fields.

Features	Spring Barley		Faba Bean	
	Earthworms			
	Individuals	Biomass	Individuals	Biomass
residues	0.095	0.030	−0.116	0.057
SOM	0.192	0.105	0.730 *	0.713 *
pH	0.120	0.013	−0.228	−0.386

* coefficient significant at $p < 0.05$.

4. Conclusions

In the current study post-harvest residues were unaffected by cropping system, but HF+ compared to HF− increased spring barely shoots residues and total post-harvest biomass and decreased weed post-harvest biomass. Experimental factors did not differentiate SOM in spring barley fields. CC, in relation to CR, increased soil pH whereas HF+ decreased it in comparison with HF−. In spring barley fields, two earthworm species were found: *Aporrectodea caliginosa* (Sav.) and *Aporrectodea rosea* (Sav.). CC and HF+ did not affect earthworm density or biomass in relation to CR and HF−. No correlation between earthworm abundance and post-harvest residues, SOM, or soil pH was noted.

In faba bean fields above-ground post-harvest residue biomass was increased by CC in relation to CR. In turn, HF+ compared with HF− increased total and faba bean shoots and roots post-harvest biomass but decreased weeds post-harvest biomass. In faba bean fields, SOM stayed at the same level regardless of experimental factors. Lower pH values were noted under CC than CR treatment, whereas HF+ did not affect it. *Aporrectodea caliginosa* (Sav.) and *Lumbricus terrestris* (L.) were recorded in faba bean fields. CC increased earthworm density and biomass in comparison with CR whereas HF+ decreased these features in relation to HF−. Positive correlations between earthworm density and biomass and SOM content were noted in faba bean fields.

Author Contributions: Conceptualization, K.T., M.J., and M.K.K.; methodology, K.T., M.J., and M.K.K.; validation, M.K.K. and P.M.; investigation, K.T. and P.M.; formal analysis, K.T., M.J., and P.M.; data curation, K.T. and P.M.; visualization, K.T., M.J., and M.K.K.; writing—original draft preparation, K.T.; writing—review and editing, M.J. and M.K.K.; funding acquisition, K.T. All authors have read and agreed to the published version of the manuscript.

Funding: This research was financially supported by the Minister of Science and Higher Education within the framework of the program entitled "Regional Initiative of Excellence" for the years 2019–2022, Project No. 010/RID/2018/19, amount of funding 12.000.000 PLN.

Conflicts of Interest: The authors declare no conflict of interest.

References

1. Blouin, M.; Hodson, M.E.; Delgado, E.A.; Baker, G.; Brussaard, L.; Butt, K.R.; Dai, J.; Dendooven, L.; Peres, G.; Tondoh, J.E.; et al. A review of earthworm impact on soil function and ecosystem services. *Eur. J. Soil Sci.* **2013**, *64*, 161–182. [CrossRef]

2. Bottinelli, N.; Henry-des-Tureaux, T.; Hallaire, V.; Mathieu, J.; Benard, Y.; Duc Tran, T.; Jouquet, P. Earthworms accelerate soil porosity turnover under watering conditions. *Geoderma* **2010**, *156*, 43–47. [CrossRef]

3. Pelosi, C.; Barot, S.; Capowiez, Y.; Hedde, M.; Vandenbulcke, F. Pesticides and earthworms. A review. *Agron. Sustain. Dev.* **2014**, *34*, 199–228. [CrossRef]

4. Jouquet, P.; Podwojewski, P.; Bottinelli, N.; Mathieu, J.; Ricoy, M.; Orange, D.; Tran, T.D.; Valentin, C. Above-ground earthworm casts affect water runoff and soil erosion in Northern Vietnam. *Catena* **2008**, *74*, 13–21. [CrossRef]

5. Winding, A.; Rønn, R.; Hendriksen, N.B. Bacteria and protozoa in soil microhabitats as affected by earthworms. *Biol. Fertil. Soils* **1997**, *24*, 133–140. [CrossRef]

6. Scheu, S.; Schlitt, N.; Tiunov, A.V.; Newington, J.E.; Jones, T.H. Effects of the presence and community composition of earthworms on microbial community functioning. *Oecologia* **2002**, *133*, 254–260. [CrossRef]

7. Jouquet, P.; Bottinelli, N.; Podwojewski, P.; Hallaire, V.; Tran Duc, T. Chemical and physical properties of earthworm casts as compared to bulk soil under a range of different land-use systems in Vietnam. *Geoderma* **2008**, *146*, 231–238. [CrossRef]

8. Vos, H.M.J.; Koopmans, G.F.; Beezemer, L.; de Goede, R.G.M.; Hiemstra, T.; van Groenigen, J.W. Large variations in readily-available phosphorus in casts of eight earthworm species are linked to cast properties. *Soil Biol. Biochem.* **2019**, *138*, 107583. [CrossRef]

9. Eriksen-Hamel, N.S.; Whalen, J.K. Impacts of earthworms on soil nutrients and plant growth in soybean and maize agroecosystems. *Agric. Ecosyst. Environ.* **2007**, *120*, 442–448. [CrossRef]

10. Fonte, S.J.; Kong, A.Y.Y.; van Kessel, C.; Hendrix, P.F.; Six, J. Influence of earthworm activity on aggregate-associated carbon and nitrogen dynamics differs with agroecosystem management. *Soil Biol. Biochem.* **2007**, *39*, 1014–1022. [CrossRef]

11. Sruthi, S.N.; Ramasamy, E.V. Enrichment of soil organic carbon by native earthworms in a patch of tropical soil, Kerala, India: First report. *Sci. Rep.* **2018**, *8*, 5784. [CrossRef]

12. Krishnamoorthy, R.V.; Vajranabhaiah, S.N. Biological activity of earthworm casts: An assessment of plant growth promotor levels in the casts. *Proc. Anim. Sci.* **1986**, *95*, 341–351. [CrossRef]

13. Muscolo, A.; Bovalo, F.; Gionfriddo, F.; Nardi, S. Earthworm humic matter produces auxin-like effects on *Daucus carota* cell growth and nitrate metabolism. *Soil Biol. Biochem.* **1999**, *31*, 1303–1311. [CrossRef]

14. Blouin, M.; Zuily-Fodil, Y.; Pham-Thi, A.T.; Laffray, D.; Reversat, G.; Pando, A.; Tondoh, J.; Lavelle, P. Belowground organism activities affect plant aboveground phenotype, inducing plant tolerance to parasites. *Ecol. Lett.* **2005**, *8*, 202–208. [CrossRef]

15. Boyer, J.; Reversat, G.; Lavelle, P.; Chabanne, A. Interactions between earthworms and plant-parasitic nematodes. *Eur. J. Soil Biol.* **2013**, *59*, 43–47. [CrossRef]

16. Stephens, P.M.; Davoren, C.W.; Doube, B.M.; Ryder, M.H. Ability of the lumbricid earthworms *Aporrectodea rosea* and *Aporrectodea trapezoides* to reduce the severity of take-all under greenhouse and field conditions. *Soil Biol. Biochem.* **1994**, *26*, 1291–1297. [CrossRef]

17. Van Groenigen, J.W.; Lubbers, I.M.; Vos, H.M.J.; Brown, G.G.; De Deyn, G.B.; Van Groenigen, K.J. Earthworms increase plant production: A meta-analysis. *Sci. Rep.* **2014**, *4*, 6365. [CrossRef] [PubMed]

18. Scheu, S. Effects of earthworms on plant growth: Patterns and perspectives. *Pedobiologia* **2003**, *47*, 846–856. [CrossRef]

19. Baldivieso-Freitas, P.; Blanco-Moreno, J.M.; Gutiérrez-López, M.; Peigné, J.; Pérez-Ferrer, A.; Trigo-Aza, D.; Sans, F.X. Earthworm abundance response to conservation agriculture practices in organic arable farming under Mediterranean climate. *Pedobiologia* **2018**, *66*, 58–64. [CrossRef]

20. Falco, L.B.; Sandler, R.; Momo, F.; Di Ciocco, C.; Saravia, L.; Coviella, C. Earthworm assemblages in different intensity of agricultural uses and their relation to edaphic variables. *PeerJ* **2015**, *3*, e979. [CrossRef]

21. Gerard, B.M.; Hay, R.K.M. The effect on earthworms of ploughing, tined cultivation, direct drilling and nitrogen in a barley monoculture system. *J. Agric. Sci.* **1979**, *93*, 147–155. [CrossRef]

22. Curry, J.P.; Byrne, D.; Schmidt, O. Intensive cultivation can drastically reduce earthworm populations in arable land. *Eur. J. Soil Biol.* **2002**, *38*, 127–130. [CrossRef]

23. Emmerling, C. Response of earthworm communities to different types of soil tillage. *Appl. Soil Ecol.* **2001**, *17*, 91–96. [CrossRef]

24. Capowiez, Y.; Cadoux, S.; Bouchant, P.; Ruy, S.; Roger-Estrade, J.; Richard, G.; Boizard, H. The effect of tillage type and cropping system on earthworm communities, macroporosity and water infiltration. *Soil Tillage Res.* **2009**, *105*, 209–216. [CrossRef]

25. Metzke, M.; Potthoff, M.; Quintern, M.; Heß, J.; Joergensen, R.G. Effect of reduced tillage systems on earthworm communities in a 6-year organic rotation. *Eur. J. Soil Biol.* **2007**, *43*, 209–215. [CrossRef]

26. Bohlen, P.J.; Edwards, C.A.; Zhang, Q.; Parmelee, R.W.; Allen, M. Indirect effects of earthworms on microbial assimilation of labile carbon. *Appl. Soil Ecol.* **2002**, *20*, 255–261. [CrossRef]

27. Parmelee, R.W.; Beare, M.H.; Cheng, W.; Hendrix, P.F.; Rider, S.J.; Crossley, D.A.; Coleman, D.C. Earthworms and enchytraeids in conventional and no-tillage agroecosystems: A biocide approach to assess their role in organic matter breakdown. *Biol. Fertil. Soils* **1990**, *10*, 1–10. [CrossRef]

28. Edwards, C.A.; Lofty, J.R. Nitrogenous fertilizers and earthworm populations in agricultural soils. *Soil Biol. Biochem.* **1982**, *14*, 515–521. [CrossRef]

29. Eriksen-Hamel, N.S.; Speratti, A.B.; Whalen, J.K.; Légère, A.; Madramootoo, C.A. Earthworm populations and growth rates related to long-term crop residue and tillage management. *Soil Tillage Res.* **2009**, *104*, 311–316. [CrossRef]

30. Edwards, C.A.; Bohlen, P.J. The effects of toxic chemicals on earthworms. *Rev. Environ. Contam. Toxicol.* **1992**, *125*, 23–100.

31. Bustos-Obregón, E.; Goicochea, R.I. Pesticide soil contamination mainly affects earthworm male reproductive parameters. *Asian J. Androl.* **2002**, *4*, 195–199.

32. Yasmin, S.; D'Souza, D. Effects of pesticides on the growth and reproduction of earthworm: A review. *Appl. Environ. Soil Sci.* **2010**, *2010*, 678360. [CrossRef]

33. Ndubuisi, S.O.; Obiezue, R.N.; Amarachi, O.; Onuoha, E. Toxicity and histopathological effect of atrazine (Herbicide) on the earthworm *Nsukkadrilus mbae* under laboratory conditions. *Anim. Res. Int.* **2010**, *7*, 1287–1293.

34. Xiao, N.; Jing, B.; Ge, F.; Liu, X. The fate of herbicide acetochlor and its toxicity to *Eisenia fetida* under laboratory conditions. *Chemosphere* **2006**, *62*, 1366–1373. [CrossRef] [PubMed]

35. Correia, F.V.; Moreira, J.C. Effects of glyphosate and 2,4-D on earthworms (*Eisenia foetida*) in laboratory tests. *Bull. Environ. Contam. Toxicol.* **2010**, *85*, 264–268. [CrossRef] [PubMed]

36. Muthukaruppan, G.; Janardhanan, S.; Vijayalakshmi, G.S. Sublethal toxicity of the herbicide butachlor on the earthworm *Perionyx sansibaricus* and its histological changes. *J. Soils Sediments* **2005**, *5*, 82–86. [CrossRef]

37. Gobi, M.; Gunasekaran, P. Effect of butachlor herbicide on earthworm *Eisenia fetida*—Its histological perspicuity. *Appl. Environ. Soil Sci.* **2010**, *2010*, 1–4. [CrossRef]

38. Singh, V.; Singh, K. Toxic Effect of herbicide 2,4-D on the earthworm *Eutyphoeus waltoni* Michaelsen. *Environ. Process.* **2015**, *2*, 251–260. [CrossRef]

39. Givaudan, N.; Wiegand, C.; Le Bot, B.; Renault, D.; Pallois, F.; Llopis, S.; Binet, F. Acclimation of earthworms to chemicals in anthropogenic landscapes, physiological mechanisms and soil ecological implications. *Soil Biol. Biochem.* **2014**, *73*, 49–58. [CrossRef]

40. Duc, G. Faba bean (*Vicia faba* L.). *Field Crop. Res.* **1997**, *53*, 99–109. [CrossRef]

41. Kasprzak, K. *Skaposzczety Glebowe, III: Rodzina; dżdżownice (Lumbricidae)*; MiIZ PAN: Warszawa, Poland, 1986.

42. Zhang, H.; Wang., J.J. *Soil pH Changes During Legume Growth and Application of Plant Material*; Southern Extension and Research Activity Information Exchange Group 6; University of Georgia: Athens, Greece, 2014.

43. Lv, J.; Dong, Y.; Dong, K.; Zhao, Q.; Yang, Z.; Chen, L. Intercropping with wheat suppressed Fusarium wilt in faba bean and modulated the composition of root exudates. *Plant Soil.* **2020**, *448*, 153–164. [CrossRef]

44. Rychcik, B.; Zawiślak, K. Effect of crop rotation and monoculture on faba bean phytosanitary and production. *Pam. Puł.* **2002**, *130*, 623–659.

45. Kotecki, A. The influence of row spacing and quantity of seeds sown on the development and yielding of Tibo cultivar field bean with terminal inflorescence Zesz. *Nauk. AR We Wrocławiu.* **1994**, *238*, 133–143.

46. Rychcik, B. Effect of herbicide and crop sequence on weed infestation of field bean. *Prog. Plant Prot.* **2004**, *44*, 1058–1060.

47. Adamiak, E.; Adamiak, J.; Stępień, A. Wpływ następstwa roślin i stosowania herbicydów na zachwaszczenie jęczmienia jarego. *Ann. Univ. Mariae Curie-Skłodowska* **2000**, *55*, 9–15.

48. Rychcik, B.; Adamiak, J.; Wójciak, H. Dynamics of the soil organic matter in crop rotation and long-term monoculture. *Plant Soil Environ.* **2006**, *52*, 12–20.

49. Yang, X.M.; Kay, B.D. Rotation and tillage effects on soil organic carbon sequestration in a typic Hapludalf in southern Ontario. *Soil Tillage Res.* **2001**, *59*, 107–114. [CrossRef]

50. Campbell, C.A.; Biederbeck, V.O.; Zentner, R.P.; Lafond, G.P. Effect of crop rotations and cultural practices on soil organic matter, microbial biomass and respiration in a thin Black Chernozem. *Can. J. Soil Sci.* **1991**, *71*, 363–376. [CrossRef]

51. Blair, N.; Crocker, G.J. Crop rotation effects on soil carbon and physical fertility of two Australian soils. *Aust. J. Soil Res.* **2000**, *38*, 71–84. [CrossRef]

52. Hickman, M.V. Long-term tillage and crop rotation effects on soil chemical and mineral properties. *J. Plant Nutr.* **2002**, *25*, 1457–1470. [CrossRef]

53. Van Beusichem, M.L. Nutrient absorption by pea plants during dinitrogen fixation. I. Comparison with nitrate nutrition. *Neth. J. Agric. Sci.* **1981**, *29*, 259–272.

54. Bolan, N.S.; Hedley, M.J.; White, R.E. Processes of soil acidification during nitrogen cycling with emphasis on legume based pastures. *Plant Soil* **1991**, *134*, 53–63. [CrossRef]

55. Tang, C.; Unkovich, M.J.; Bowden, J.W. Factors affecting soil acidification under legumes. III. Acid production by N2-fixing legumes as influenced by nitrate supply. *New Phytol.* **1999**, *143*, 513–521. [CrossRef]

56. Lee, B. Farming brings acid soils. *Rural Res.* **1980**, *106*, 4–9.

57. Williams, C.H. Soil acidification under clover pasture. *Aust. J. Exp. Agric.* **1980**, *20*, 561–567. [CrossRef]

58. Valchovski, H. Effect of erosion on biodiversity of earthworms (*Lumbricidae*) in agroecosystems of Suhodol (Sofia province). *Semin. Ecol.* **2012**, 182–189.

59. Fragoso, C.; Brown, G.G.; Patrón, J.C.; Blanchart, E.; Lavelle, P.; Pashanasi, B.; Senapati, B.; Kumar, T. Agricultural intensification, soil biodiversity and agroecosystem function in the tropics: The role of earthworms. *Appl. Soil Ecol.* **1997**, *6*, 17–35. [CrossRef]

60. Paoletti, M.G. The role of earthworms for assessment of sustainability and as bioindicators. *Agric. Ecosyst. Environ.* **1999**, *74*, 137–155. [CrossRef]

61. Curry, J.P.; Doherty, P.; Purvis, G.; Schmidt, O. Relationships between earthworm populations and management intensity in cattle-grazed pastures in Ireland. *Appl. Soil Ecol.* **2008**, *39*, 58–64. [CrossRef]

62. Edwards, C.A. Earthworm ecology in cultivated soils. In *Earthworm Ecology*; Springer: Dordrecht, The Netherlands, 1983; pp. 123–137.

63. Bart, S.; Amossé, J.; Lowe, C.N.; Mougin, C.; Péry, A.R.R.; Pelosi, C. *Aporrectodea caliginosa*, a relevant earthworm species for a posteriori pesticide risk assessment: Current knowledge and recommendations for culture and experimental design. *Environ. Sci. Pollut. Res.* **2018**, *25*, 33867–33881. [CrossRef]

64. Feledyn-Szewczyk, B.; Radzikowski, P.; Stalenga, J.; Matyka, M. Comparison of the effect of perennial energy crops and arable crops on earthworm populations. *Agronomy* **2019**, *9*, 675. [CrossRef]

65. Holmstrup, M.; Overgaard, J. Freeze tolerance in *Aporrectodea caliginosa* and other earthworms from Finland. *Cryobiology* **2007**, *55*, 80–86. [CrossRef] [PubMed]

66. Zorn, M.I.; Van Gestel, C.A.M.; Morrien, E.; Wagenaar, M.; Eijsackers, H. Flooding responses of three earthworm species, *Allolobophora chlorotica*, *Aporrectodea caliginosa* and *Lumbricus rubellus*, in a laboratory-controlled environment. *Soil Biol. Biochem.* **2008**, *40*, 587–593. [CrossRef]

67. Nuutinen, V. Earthworm community response to tillage and residue management on different soil types in southern Finland. *Soil Tillage Res.* **1992**, *23*, 221–239. [CrossRef]

68. Holter, P. Effect of earthworms on the disappearance rate of cattle droppings. In *Earthworm Ecology*; Springer: Dordrecht, The Netherlands, 1983; pp. 49–57.

69. Ivask, M.; Kuu, A.; Sizov, E. Abundance of earthworm species in Estonian arable soils. *Eur. J. Soil Biol.* **2007**, *43*, 39–42. [CrossRef]

70. Pfiffner, L.; Luka, H. Earthworm populations in two low-input cereal farming systems. *Appl. Soil Ecol.* **2007**, *37*, 184–191. [CrossRef]

71. Jastrzebska, M.; Kostrzewska, M.K.; Makowski, P.; Treder, K.; Marks, M. Effects of ash and bone phosphorus biofertilizers on *Bacillus megaterium* counts and select biological and physical soil properties. *Polish J. Environ. Stud.* **2015**, *24*, 1603–1609. [CrossRef]

72. Orzech, K.; Załuski, D. Chemical properties of soil and occurrence of earthworms in soil in response to soil compaction and different soil tillage in cereals. *J. Elem.* **2020**, *25*, 153–168. [CrossRef]

73. Jastrzębska, M.; Kostrzewska, M.; Makowski, P.; Treder, K.; Jastrzębski, W. Granulated phosphorus fertilizer made of ash from biomass combustion and bones with addition of *Bacillus megaterium* in the field assessment. Part 3. Impact on selected properties of soil environment of winter wheat. *Przem. Chem.* **2017**, *96*, 2180–2183.

74. Orzech, K.; Załuski, D. Effect of soil compaction and different soil tillage systems on chemical properties of soil and presence of earthworms in winter oilseed rape fields. *J. Elem.* **2020**, *25*, 413–429.

75. Badawy, M.E.I.; Kenawy, A.; El-Aswad, A.F. Toxicity assessment of buprofezin, lufenuron, and triflumuron to the earthworm *Aporrectodea caliginosa*. *Int. J. Zool.* **2013**, *2013*, 174523. [CrossRef]

76. Mosleh, Y.Y.; Ismail, S.M.M.; Ahmed, M.T.; Ahmed, Y.M. Comparative toxicity and biochemical responses of certain pesticides to the mature earthworm *Aporrectodea caliginosa* under laboratory conditions. *Environ. Toxicol.* **2003**, *18*, 338–346. [CrossRef] [PubMed]

77. Gaupp-Berghausen, M.; Hofer, M.; Rewald, B.; Zaller, J.G. Glyphosate-based herbicides reduce the activity and reproduction of earthworms and lead to increased soil nutrient concentrations. *Sci. Rep.* **2015**, *5*, 12886. [CrossRef]

78. Dittbrenner, N.; Triebskorn, R.; Moser, I.; Capowiez, Y. Physiological and behavioural effects of imidacloprid on two ecologically relevant earthworm species (*Lumbricus terrestris* and *Aporrectodea caliginosa*). *Ecotoxicology* **2010**, *19*, 1567–1573. [CrossRef] [PubMed]

79. Capowiez, Y.; Rault, M.; Costagliola, G.; Mazzia, C. Lethal and sublethal effects of imidacloprid on two earthworm species (*Aporrectodea nocturna* and *Allolobophora icterica*). *Biol. Fertil. Soils* **2005**, *41*, 135–143. [CrossRef]

80. Hodge, S.; Webster, K.M.; Booth, L.; Hepplethwaite, V.; O'Halloran, K. Non-avoidance of organophosphate insecticides by the earthworm *Aporrectodea caliginosa* (*Lumbricidae*). *Soil Biol. Biochem.* **2000**, *32*, 425–428. [CrossRef]

81. López-Hernández, D. Earthworm populations in savannas of the Orinoco basin. A review of studies in long-term agricultural-managed and protected ecosystems. *Agriculture* **2012**, *2*, 87–108. [CrossRef]

82. Kladivko, E.J. *Earthworms and Crop Management.*; Purdue University: West Lafayette, IN, USA, 1993.

83. Fonte, S.J.; Winsome, T.; Six, J. Earthworm populations in relation to soil organic matter dynamics and management in California tomato cropping systems. *Appl. Soil Ecol.* **2009**, *41*, 206–214. [CrossRef]

84. Haynes, R.J.; Dominy, C.S.; Graham, M.H. Effect of agricultural land use on soil organic matter status and the composition of earthworm communities in Kwazulu-Natal, South Africa. *Agric. Ecosyst. Environ.* **2003**, *95*, 453–464. [CrossRef]

Communication

Phosphorus Fertilizers From Sewage Sludge Ash and Animal Blood Have No Effect on Earthworms

Magdalena Jastrzębska *, Marta K. Kostrzewska and Kinga Treder

Department of Agroecosystems, Faculty of Environmental Management and Agriculture, University of Warmia and Mazury in Olsztyn, Plac Łódzki 3, 10-718 Olsztyn, Poland; marta.kostrzewska@uwm.edu.pl (M.K.K.); kinga.treder@uwm.edu.pl (K.T.)
* Correspondence: jama@uwm.edu.pl; Tel.: +48-89-523-4829

Received: 9 March 2020; Accepted: 5 April 2020; Published: 7 April 2020

Abstract: Soil invertebrates are crucial for agroecosystem functioning yet sensitive to agricultural practices, including fertilization. Considering the postulates of circular phosphorus economy, the use of fertilizers from secondary raw materials is likely to return and increase and may even become obligatory. The effects of recycled fertilizers on soil fauna communities, however, remain poorly understood. In this paper, the effect of phosphorus fertilizer (RecF) and biofertilizer (RecB) from sewage sludge ash and dried animal (porcine) blood on earthworm's occurrence in soil is discussed. RecB is RecF activated by phosphorus-solubilizing bacteria, *Bacillus megaterium*. Waste-based fertilizers were assessed in field experiments against commercial superphosphate and no P fertilization. Three levels of P doses were established (17.6, 26.4, and 35.2 kg P ha^{-1}). Earthworms were collected after the test crop harvest (spring or winter wheat). In the experiments two earthworm species, *Aporrectodea caliginosa* and *Aporrectodea rosea*, were identified. A large proportion of juvenile individuals were recorded in 2017. The recycled fertilizers used in the experiments used in recommended doses, similarly to superphosphate, did not alter the density, biomass, species composition, and structure of earthworms. Further long-term field research is recommended.

Keywords: *Lumbricidae*; *Aporrectodea caliginosa*; *Aporrectodea rosea*; phosphorus fertilizers; phosphorus-solubilizing microorganisms; renewable resources; heavy metals; Luvisols; wheat

---

1. Introduction

Earthworms (*Lumbricidae*) are listed among the most important soil-dwelling invertebrates [1]. They constitute a major component of soil fauna communities in most ecosystems [2]. The role of earthworms in soil fertility has been known for over a century [2]. So far, a great number of studies have been undertaken which highlight direct and indirect effects of their activity on biotic and abiotic soil properties, and, consequently, plant productivity. Due to their services, earthworms are referred to as ecosystem engineers [3,4] and indicators of biological soil health [5,6].

The occurrence, distribution, and abundance of earthworms can be affected by a range of environmental factors, including climate, soil conditions, food sources, metal concentration, and predator pressure [5]. In addition, in agroecosystems, agricultural practices such as irrigation, tillage, lime application, fertilizer and pesticide use, drainage, crop rotation, and cover crops influence earthworm abundance and activity [7] because they change one or more of the factors listed above [5,8].

Despite potential soil pollution [9], increased use of inorganic fertilizers to enhance crop yields is a common practice in modern agriculture. Both beneficial and harmful effects of inorganic fertilizers on earthworm populations have been observed [10]. The positive effect is believed to be an indirect consequence of increased crop biomass production and the resulting increase in organic residues [11]. On the other hand, the toxic effects of inorganic fertilizers on earthworms, especially upon direct contact, have been reported [12,13].

Modern European agriculture faces a shortage of primary phosphorus (P) sources. Phosphate rock was included in the EU list of critical resources in 2014 [14]. A circular P economy, including recycling, seems to be a necessity in this part of the world. Inorganic and organic waste are often a source of nutrients in fertilizers [15,16]. As has been proved in numerous scientific centers, phosphate rocks can be replaced with P-rich secondary raw materials [17–19]. Municipal and industrial byproducts such as sewage sludge ash (SSA), animal bones, and blood may constitute the basis for alternative fertilizers [19]. An innovative approach, initiated to activate P from raw material, is the inclusion of phosphorus-solubilizing microbes (PSM) into waste-based preparations [20]. The use of recycled fertilizers is expected not only to provide satisfactory yields in terms of quantity [21,22] and quality, but also not to cause negative changes in the soil environment. Concerning the latter, it should be taken into account that the introduction of nutrient carrier and PSM to the soil could alter soil properties both directly (nutrient content and availability, pH, possible presence of toxic elements) and indirectly (e.g., through microbial activity modification or plant growth stimulation) [23]. Changes in habitat conditions could affect earthworm populations. It is also crucial to be aware that the consequences of recycled fertilizer use, while being invisible in the short term, may lead to significant environmental changes in the long term [24,25].

The aim of this research has been to determine the impact of the fertilizers produced from SSA and animal blood on earthworm occurrence in the soil. The recycled fertilizer (RecF) and biofertilizer (RecB), i.e., RecF activated by *Bacillus megaterium* bacteria (PSM) were assessed against superphosphate, a commercial phosphorus fertilizer. It was hypothesized that the impact of the recycled fertilizers on soil earthworms would be similar or more favorable/less harmful than that of the traditional P fertilizer.

2. Materials and Methods

2.1. Fertilizers

In field experiments, the recycled P fertilizer (RecF) and biofertilizer (RecB) were compared to a commercial fertilizer superphosphate (SP). These preparations were manufactured from sewage sludge ash (ash from the incineration of sewage sludge biomass from wastewater treatment; SSA) and dried animal (porcine) blood. During RecB production, raw material (SSA + blood) was biologically activated by phosphorus-solubilizing bacteria, *Bacillus megaterium*. Both products were in the form of granules.

RecF and RecB were produced at the Institute of New Chemical Syntheses in Puławy (Poland), according to a concept developed at the Wrocław University of Science and Technology (Wrocław, Poland). The SSA originated from the Municipal Wastewater Treatment Plant 'Łyna' in Olsztyn (Poland), and dried blood was obtained from the meat processing industry. The bacteria strains were obtained from the Polish Collection of Microorganisms at the Institute of Immunology and Experimental Therapy of the Polish Academy of Sciences in Wrocław (Poland). The elemental composition of the recycled fertilizers is presented in Table 1. The production process was described by Rolewicz et al. [26].

Table 1. Elemental composition of the recycled fertilizers.

Element	Unit	2016		2017	
		RecF	RecB	RecF	RecB
P		8.68	9.55	5.40	4.95
N		2.89	2.87	3.44	3.15
K		1.09	1.16	0.62	0.67
C	% mass.	13.4	14.6	14.2	12.3
Mg		1.54	1.70	0.79	0.78
S		0.56	0.56	0.47	0.40
C		12.5	13.9	16.5	18.1
Fe		26.9	29.0	11.4	11.3
Al	$g\ kg^{-1}$	23.7	25.5	11.3	12.1
Zn		3.14	3.29	1.09	0.99
As		31.4	20.0	15.5	20.5
Cd		<0.01	0.345	0.660	0.742
Cr		54.7	62.9	63.9	59.1
Cu		778	850	334	334
Ni		54.8	62.6	28.5	21.2
Pb	$mg\ kg^{-1}$	19.9	21.8	0.920	4.53
B		71.3	74.1	41.1	57.6
Ba		349	382	162	168
Co		14.0	16.2	5.24	4.24
Mn		562	609	299	437
Mo		35.3	23.7	9.25	13.9

According to the Department of Advanced Material Technologies of the Wrocław University of Science and Technology (Wrocław, Poland).

Superphosphate Fosdar[TM] 40 (Gdańsk Phosphorus Fertilizer Plant 'Fosfory' Sp. z o.o., Gdańsk, Poland) was purchased on the market. This P fertilizer contains 17.6% P, 7.15% Ca, 2.00% S, and microelements (B, Co, Cu, Fe, Mn, Mo, and Zn), according to the commercial information provided on the label.

2.2. Soil and Meteorological Conditions

Three field experiments with spring (2016, 2017) or winter (2017; sown in autumn 2016) common wheat (*Triticum aestivum* ssp. *vulgare* MacKey) were conducted. In each experiment, the soil on which wheat was grown met the requirements of the species (Table 2) and was within the range of soils preferred by earthworms [27].

Table 2. Soil characteristics before the start of the experiments.

Experiment	Soil Type	Soil Texture	pH in KCl	Total, $g\ kg^{-1}$				
				C	N	P	K	Mg
Spring wheat 2016	Luvisols [1]	sandy clay loam	6.28	8.53	1.42	0.61	2.98	2.02
Spring wheat 2017	Luvisols	sandy clay loam	6.23	8.48	1.34	0.60	3.14	1.94
Winter wheat 2017	Luvisols	sandy loam	4.98	6.48	1.01	0.49	2.95	1.88

[1] According to World reference base for soil resources 2014 [28].

Meteorological conditions in the period of one month before earthworm sampling are presented in Table 3. In both growing seasons, fairly heavy rainfall and moderate temperatures in July and early August could have stimulated earthworm activity at the time of earthworm sampling [29,30].

Table 3. Atmospheric precipitation and air temperature during the study period according to the Meteorological Station in Bałcyny, Poland.

Year	Month	Atmospheric Precipitation (mm)				Air Temperature ($^\circ$C)			
		Period of Ten Days			Total	Period of Ten Days			Average
		1st	2nd	3rd		1st	2nd	3rd	
2016	July	39.6	34.0	65.0	138.6	17.5	18.1	19.9	18.5
	August	54.5	10.4	7.0	71.9	17.7	15.6	19.3	17.6
2017	July	29.4	20.7	56.0	106.1	16.0	17.2	18.5	17.3
	August	31.6	11.7	11.5	54.8	20.8	19.2	16.3	18.7

2.3. Experimental Design and Agronomic Management

In the field experiments, RecF and RecB were assessed against SP and no phosphorus (No P) treatments. In addition, three different P levels were established: (1) 17.6, (2) 26.4, and (3) 35.2 kg P ha^{-1}; therefore, finally, ten treatments of P fertilization were compared (Table 4).

Table 4. Fertilization treatments compared in the experiments.

Treatment Symbol	Fertilizer	P Dose, kg P ha^{-1}
No P	without phosphorus fertilizer	0
SP$_1$	superphosphate	17.6
SP$_2$		26.4
SP$_3$		35.2
RecF$_1$	fertilizer from sewage sludge ash and	17.6
RecF$_2$	dried animal blood	26.4
RecF$_3$		35.2
RecB$_1$	biofertilizer from sewage sludge ash	17.6
RecB$_2$	and dried animal blood	26.4
RecB$_3$		35.2

Phosphorus fertilizers were applied before the sowing of wheat. They were manually scattered on the soil surface and then mixed with the soil by harrowing. Other basic agrotechnical data for the experiments are presented in Table 5.

Table 5. Basic agricultural data for the experiments.

Item	Experiment		
	Spring Wheat 2016	Spring Wheat 2017	Winter Wheat 2017
Wheatcultivar	Monsun	Monsun	Julius
Previous crop	winter rape	spring wheat	winter wheat
Soil tillagesystem	plough tillage	plough tillage	plough tillage
Fertilization			
–K$_2$O [1],kg ha^{-1}	100	100	100
–N [2], kg ha^{-1}	130	110	150
Plantprotection			
–Herbicides	florasulam + 2,4-D (29 May)	florasulam + 2,4-D (22 May)	florasulam + 2,4-D (13 May)
–Fungicides	thiophanate-methyl +tetraconazole (9 June)azoxystrobin (8 July)	thiophanate-methyl + tetraconazole (6 June) azoxystrobin + (propiconazole + cyproconazole) (28 June)	fenpropimorph + epoxiconazole + metrafenone (16 May)fluxapyroxad + pyraclostrobin + epoxiconazole (8 June)
–Insecticides	lambda-cyhalothrin(6 June)	deltamethrin (6 June)	deltamethrin (6 June)
–Growthregulators	–	–	trinexapac-ethyl (16 May)
Sowing date	21 April 2016	20 April 2017	4 October 2016
Harvest date	12 August 2016	18 August 2017	4 August 2017

[1] potassium chloride, [2] ammonium sulphate, – not applied.

Experiments were established in a randomized block design. In each experiment, particular experimental treatments were performed in four replications (plots) (Figure S1). The area of a single experimental plot was 20 m^2.

2.4. Earthworm Sampling and Identification

Earthworms were harvested mechanically 2–3 days after the wheat harvest. Soil columns with a surface area of 0.0625 m^2 (0.25 m × 0.25 m) and a depth of 0.4 m were dug out of each plot, then crushed and passed through a sieve, and individuals of *Lumbricidae* were collected. Afterwards, the earthworms were transported to the laboratory, where they were washed, counted, and weighed. Anaesthetized in a 30% ethanol (Czempur, Piekary Śląskie, Poland) solution, earthworms were preserved in a 4% formalin (Czempur, Piekary Śląskie, Poland) and 75% ethanol solution for the subsequent analysis of the species composition. The earthworms were sorted into adults and juvenile forms. The adult individuals were further classified into species using an identification key to soil-dwelling oligochaetes [31]. The species composition, number and biomass of earthworms in the 0–0.4 m soil layer were expressed per 1 m^2 of plot area.

2.5. Statistical Analysis

The normality of variable distribution was checked using the Shapiro–Wilk W-test, and the homogeneity of variance was checked using Levene's test. Since the assumptions of the analysis of variance were not met, the results were processed by the alternative nonparametric Kruskal–Wallis test. The calculations were performed using Statistica 12.0 software [32].

3. Results and Discussion

Both in 2016 and 2017, thermal and rain conditions in July and early August promoted earthworm presence in the 0–0.4 m soil layer. Having found convenient habitat moisture at this level of the soil profile, the individuals of *Lumbricidae* did not enter into diapause or migrate deeper into the soil seeking better conditions [30]. The density of earthworms found in the studied soil columns ranged from 6 to 44 individuals and the biomass from 1.1 to 21.5 g per m^2 (Table 6). These values are similar to those presented by Tiwari [33] from a sandy loam Oxisol in India, but smaller than the values reported by other authors from different arable soils in Poland [4] and Slovakia [34]. The abovementioned differences may have been caused by different timing of sampling, which did not correspond to the periods of the highest earthworm activity (spring and autumn) indicated in the literature [4,34]. In 2016, the average earthworm biomass was relatively higher than in 2017 due to a greater share of adult individuals in the community.

Table 6. Earthworm density (no. m^{-2}) and biomass (g m^{-2}) in the 0–0.4 m layer of soil under wheat (averages from four replications/plots).

P Treatment	Spring Wheat 2016		Spring Wheat 2017		Winter Wheat 2017	
	Earthworm					
	Density	Biomass	Density	Biomass	Density	Biomass
No P	22	16.3	18	7.8	6	1.1
SP$_1$	18	9.8	24	5.4	22	7.6
SP$_2$	26	21.5	32	12.0	12	7.3
SP$_3$	18	9.4	20	8.0	16	8.5
RecF$_1$	26	15.0	12	3.2	20	8.1
RecF$_2$	8	6.7	16	3.5	20	11.0
RecF$_3$	12	6.1	12	3.6	12	6.6
RecB$_1$	14	11.4	16	6.7	10	3.8
RecB$_2$	16	9.1	6	1.6	14	8.0
RecB$_3$	28	16.4	44	11.2	8	4.9

No significant differences between treatments according to the Kruskal–Wallis test at $p \leq 0.05$.

In all experiments, only two earthworm species were identified, i.e., *Aporrectodea caliginosa* and *Aporrectodea rosea* (Figure 1), which is hardly surprising. These species are among the most common in Poland [31] and Europe [35], and they were the only ones recorded by Kanianska et al. [34] in some study sites in Slovakia. In 2016, mainly adult earthworms were noted, and on average, *A. caliginosa* and *A. rosea* occurred in similar proportions (42% and 39%, respectively). In 2017, among the earthworm individuals found after spring wheat harvest, juvenile forms dominated, often constituting 100% of the community. Adults were found sporadically. A large proportion of juvenile forms (mostly over 50%) were also recorded in the soil after the winter wheat harvest. In this experiment, *A. rosea* was predominant. A high number of juvenile individuals is often thought to be an indicator of suitable conditions for earthworm development [29,36]. A dominance of juvenile forms over adult earthworms has also been noticed by other authors [4,34].

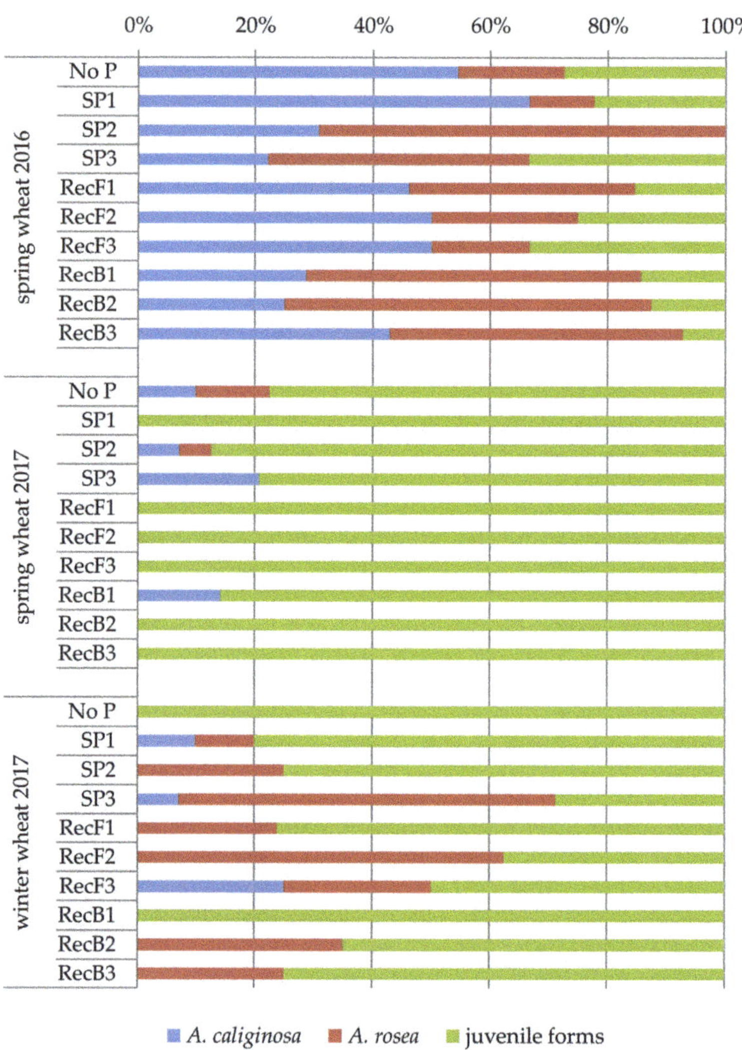

Figure 1. Species composition and structure of earthworms (based on the density of individuals). No significant differences between treatments according to the Kruskal–Wallis test at $p \leq 0.05$.

In none of the conducted experiments did the earthworm density and biomass depend on the type of P fertilizers used or their doses (Table 6). Moreover, earthworm abundance (density and biomass) under no P treatment did not differ from that under fertilizers. In addition, no evident link between the species composition and structure of earthworms and the applied P fertilization was observed (Figure 1).

To compare, in the study by Tiwari [33] conducted in an Oxisol (India), the single superphosphate applied at P dose of 25 kg ha^{-1} did not change the earthworm density and biomass in comparison to control treatment (no fertilizer). An increase in the number and biomass of earthworms with the addition of superphosphate to pastures in Australia and New Zealand was reported [37]; however, the authors argued that P fertilizer led to an increase in plant production in these ecosystems and, hence, available food. In contrast, in other studies [34,38,39], a negative relationship between earthworm biomass and P content in soil was found. Some authors proved that inorganic fertilizers, including superphosphate, can be toxic to earthworms upon direct contact [12,13].

In the current study, the SSA is the main raw material for the fertilizers produced, and one that may raise concerns about the heavy metal presence [18]. The issue of toxic element occurrence is key since Khan et al. [40], based on a pot experiment, claimed that the high content of heavy metals in the tested fiber and chemical industry sludge ashes was the reason for the decrease in the number of adults, juveniles, cocoons, and fresh weight of the earthworm *Pheretima posthuma* found four months after the waste application. Using animal blood as a fertilizer for organic farming [41,42] and a fertilizer binder [43] was recommended. The content of potentially toxic elements in fertilizers tested in the current study was low (Table 1), and the fertilizer doses used were not excessive. According to other research, metals such as copper (Cu), zinc (Zn), and iron (Fe), which are contained in RecF and RecB fertilizers, may also be toxic to earthworms [13,44,45], although they play the role of microelements for plants. Neuhauser et al. [44] proved that Cu and Zn were more toxic to *Eisenia fetida* than cadmium (Cd) and lead (Pb). Toxicity of aluminum (Al) to earthworms was reported as well [46]. Additional reflections (and caution) should also be prompted by studies on long-term use of sewage sludge documenting the negative impact of metal accumulation in the soil on soil microorganisms [24,25,47].

To date, only a few studies have examined the effect of SSA-based fertilizers on earthworms. Rastetter et al. [48] ecotoxicologically analyzed three crystallization products and five ash products of recovered phosphate-containing materials, obtained from treated sewage sludge, sludge liquors or sludge ashes from municipal wastewater treatment plants in Europe. The phosphate recyclates were compared with a conventional phosphate fertilizer (triple superphosphate). The avoidance test with the earthworm *Eisenia fetida* was used to determine the effects of chemicals on behavior of earthworms. The authors concluded that relevant agronomical application amounts of all phosphate recyclates and triple superphosphate might not have an acute toxic effect on the soil invertebrates. In contrast to endogeic species found in the current study, *E. fetida* is epigeic, and some research has suggested that the sensitivity of ecologically different earthworm species to chemicals/pollutants may vary [49,50]. The earlier field studies by Jastrzębska et al. [23,51–53] showed that suspension and granular fertilizers from SSA and/or animal bones with a low content of toxic elements and applied in recommended doses did not alter the abundance (density and biomass), species composition, and structure of soil earthworms. In the cited studies, only endogeic species were found, both in fertilized and nonfertilized soil. The current study is in line with the above results. It is also worth highlighting that the peculiar impact of PSM included in biofertilizer on earthworms was not noticed. The same results were obtained by Jastrzębska et al. [53] when fertilizer and biofertilizer from SSA and animal bones were compared. It can thus be concluded that PSM introduced into the soil in the amounts required for biofertilizers do not significantly alter the earthworm habitat conditions.

In the presented experiments, chemical plant protection was used. This may create the assumption that pesticides affected earthworms and masked the effects of fertilizers. However, in the earlier study with SSA-based suspension fertilizer, Jastrzębska et al. [23] did not observe the effect of pesticides (applied at recommended doses) on earthworms, nor the interaction between phosphorus fertilizations and plant protection (no plant protection vs. chemical plant protection). Considering the abovementioned results, we believe that this phenomenon did not occur in the presented study either.

4. Conclusions

Recycled fertilizers produced from secondary raw materials, such as sewage sludge ash with a low content of toxic elements and dried animal blood, applied in reasonable doses, similarly to superphosphate, did not pose a threat to earthworms. The impact on these organisms is not a limitation to their use. However, taking into account the potential toxicity of waste, relevant studies preceding the recommendation of each new recyclate-based product and long-term field ones are postulated.

Supplementary Materials: The following are available online at http://www.mdpi.com/2073-4395/10/4/525/s1, Figure S1: Scheme of experimental design.

Author Contributions: Conceptualization, M.J.; methodology, M.J., M.K.K., and K.T.; formal analysis, M.J.; investigation, K.T., M.J., and M.K.K.; resources, M.J. and K.T.; data curation, M.J.; writing—original draft preparation, M.J.; writing—review and editing, M.K.K.; visualization, M.J.; funding acquisition, M.J., M.K.K., and K.T. All authors have read and agreed to the published version of the manuscript.

Funding: This research was funded by the National Centre for Research and Development, Poland, grant number PBS 2/A1/11/2013.

Acknowledgments: The Institute of New Chemical Synthesis in Puławy is highly acknowledged for providing fertilizers for field experiments and Agnieszka Saeid from the Wrocław University of Science and Technology for the information on the fertilizers' chemical composition.

Conflicts of Interest: The authors declare no conflict of interest.

References

1. Van Groenigen, J.W.; Lubbers, I.M.; Vos, H.M.J.; Brown, G.G.; De Deyn, G.B.; Van Groenigen, K.J. Earthworms increase plant production: A meta-analysis. *Sci. Rep.* **2014**, *4*, 6365. [CrossRef] [PubMed]
2. Bhadauria, T.; Saxena, K.G. Role of earthworms in soil fertility maintenance through the production of biogenic structures. *Appl. Environ. Soil Sci.* **2010**, *2010*. [CrossRef]
3. Johnston, A.S.A.; Sibly, R.M.; Hodson, M.E.; Alvarez, T.; Thorbek, P. Effects of agricultural management practices on earthworm populations and crop yield: Validation and application of a mechanistic modelling approach. *J. Appl. Ecol.* **2015**, *52*, 1334–1342. [CrossRef]
4. Feledyn-Szewczyk, B.; Radzikowski, P.; Stalenga, J.; Matyka, M. Comparison of the effect of perennial energy crops and arable crops on earthworm populations. *Agronomy* **2019**, *9*, 675. [CrossRef]
5. Ashworth, A.J.; Allen, F.L.; Tyler, D.D.; Pote, D.H.; Shipitalo, M.J. Earthworm populations are affected from long-term crop sequences and bio-Covers under no-Tillage. *Pedobiologia* **2017**, *60*, 27–33. [CrossRef]
6. Doran, J.W.; Zeiss, M.R. Soil health and sustainability: Managing the biotic component of soil quality. *Appl. Soil Ecol.* **2000**, *15*, 3–11. [CrossRef]
7. Edwards, C.A.; Bohlen, P.J. *Biology and Ecology of Earthworms*, 3rd ed.; Chapman & Hall: London, UK, 1996; p. 426.
8. Baldivieso-Freitas, P.; Blanco-Moreno, J.M.; Gutiérrez-López, M.; Peigné, J.; Pérez-Ferrer, A.; Trigo-Aza, D.; Sans, F.X. Earthworm abundance response to conservation agriculture practices in organic arable farming under Mediterranean climate. *Pedobiologia* **2018**, *66*, 58–64. [CrossRef]
9. Radziemska, M.; Bęś, A.; Gusiatin, Z.M.; Majewski, G.; Mazur, Z.; Bilgin, A.; Jaskulska, I.; Brtnický, M. Immobilization of Potentially Toxic Elements (PTE) by Mineral-Based Amendments: Remediation of Contaminated Soils in Post-Industrial Sites. *Minerals* **2020**, *10*, 87. [CrossRef]
10. Lalthanzara, H.; Ramanujam, S.N. Effect of fertilizer (NPK) on earthworm population in the agro-Forestry system of Mizoram India. *Sci. Vis.* **2010**, *10*, 159–167.
11. Marhan, S.; Scheu, S. The influence of mineral and organic fertilisers on the growth of the endogeic earthworm *Octolasion tyrtaeum* (Savigny). *Pedobiologia* **2005**, *49*, 239–249. [CrossRef]
12. Abbiramy, K.S.K.; Ross, P.R.; Paramanandham, J.P. Assessment of acute toxicity of superphosphate to Eisenia foetida using paper contact method. *Asian J. Plant Sci. Res.* **2013**, *3*, 112–115.
13. Shruthi, N.; Biradar, A.P.; Muzammil, S. Toxic effect of inorganic fertilizers to earthworms (*Eudrilus eugeniae*). *J. Entomol. Zool. Stud.* **2017**, *5*, 1135–1137.
14. European Commission. Report on critical raw materials for the EU. In *Report of the Ad-Hoc Working Group on Defining Critical Raw Materials*; EC: Brussels, Belgium, 2014; p. 41.

15. Romiński, M.; Michalak, P.; Jaskulski, D.; Jaskulska, I. Technology for manufacturing granulated magnesium fertilizers with the use of wastes from chlorine production. *Przem. Chem.* **2011**, *90*, 628–630.

16. Jaskulski, D.; Jaskulska, I. Possibility of using waste from the polyvinyl chloride production process for plant fertilization. *Pol. J. Environ. Stud.* **2011**, *20*, 351–354.

17. Smol, M. The importance of sustainable phosphorus management in the circular economy (CE) model: The Polish case study. *J. Mater. Cycl. Waste Manag.* **2019**, *21*, 227–238. [CrossRef]

18. Herzel, H.; Krüger, O.; Hermann, L.; Adam, C. Sewage sludge ash-A promising secondary phosphorus source for fertilizer production. *Sci. Total Environ.* **2016**, *542*, 1136–1143. [CrossRef]

19. Saeid, A.; Wyciszkiewicz, M.; Jastrzębska, M.; Chojnacka, K.; Górecki, H. A concept of production of new generation of phosphorus-Containing biofertilizers. BioFertP project. *Przem. Chem.* **2015**, *94*, 361–365.

20. Saeid, A.; Labuda, M.; Chojnacka, K.; Górecki, H. Use of microorganism in production of phosphorus fertilizers. *Przem. Chem.* **2012**, *91*, 956–958.

21. Jastrzębska, M.; Kostrzewska, M.K.; Saeid, A.; Treder, K.; Makowski, P.; Jastrzębski, W.P.; Okorski, A. Granulated phosphorus fertilizer made of ash from biomass combustion and bones with addition of *Bacillus megaterium* in the field assessment. Part 1. Impact on yielding and sanitary condition of winter wheat. *Przem. Chem.* **2017**, *96*, 2168–2174.

22. Severin, M.; Breuer, J.; Rex, M.; Stemann, J.; Adam, C.; Van den Weghe, H.; Kücke, M. Phosphate fertilizer value of heat treated sewage sludge ash. *Plant Soil Environ.* **2014**, *60*, 555–561. [CrossRef]

23. Jastrzębska, M.; Kostrzewska, M.K. Using an environment-friendly fertiliser from sewage sludge ash with the addition of *Bacillus megaterium*. *Minerals* **2019**, *9*, 423. [CrossRef]

24. McGrath, S.P.; Chaudri, A.M.; Giller, K.E. Long-term effects of metals in sewage sludge on soils, microorganisms and plants. *J. Ind. Microbiol.* **1995**, *14*, 94–104. [CrossRef] [PubMed]

25. Charlton, A.; Sakrabani, R.; Tyrrel, S.; Rivas Casado, M.; McGrath, S.P.; Crooks, B.; Cooper, P.; Campbell, C.D. Long-term impact of sewage sludge application on soil microbial biomass: An evaluation using meta-analysis. *Environ. Pollut.* **2016**, *219*, 1021–1035. [CrossRef] [PubMed]

26. Rolewicz, M.; Rusek, P.; Borowik, K. Obtaining of granular fertilizers based on ashes from combustion of waste residues and ground bones using phosphorous solubilization by bacteria *Bacillus megaterium*. *J. Environ. Manag.* **2018**, *216*, 128–132. [CrossRef]

27. Pfiffner, L. Earthworms—architects of fertile soils. Their significance and recommendations for their promotion in agriculture. In *Technical Guide on Earthworms*; Research Institute of Organic Agriculture: Frick, Switzerland, 2014.

28. Food and Agriculture Organization of the United Nations. World reference base for soil resources 2014. In *International Soil Classification System for Naming Soils and Creating Legends for Soil Maps*; Food and Agriculture Organization of the United Nations: Rome, Italy, 2014.

29. Ivask, M.; Kuu, A.; Truu, M.; Truu, J. The effect of soil type and soil moisture on earthworm communities. *Agric. Sci.* **2006**, *17*, 7–11.

30. Singh, J.; Schädler, M.; Demetrio, W.; Brown, G.G.; Eisenhauer, N. Climate change effects on earthworms-A review. *Soil Org.* **2019**, *91*, 114.

31. Kasprzak, K. *Soil Oligochaeta. 3. Family: Earthworms (Lumbricidae)*; Państwowe Wydawnictwo Naukowe: Warsaw, Poland, 1986; p. 186.

32. *Statistica (Data Analysis Software System)*, version 12; StatSoft: Tulsa, OK, USA, 2014.

33. Tiwari, S.C. Effects of organic manure and NPK fertilization on earthworm activity in an Oxisol. *Biol. Fertil. Soil* **1993**, *16*, 293–295. [CrossRef]

34. Kanianska, R.; Jad'ud'ová, J.; Makovníková, J.; Kizeková, M. Assessment of relationships between earthworms and soil abiotic and biotic factors as a tool in sustainable agricultural. *Sustainability* **2016**, *8*, 906. [CrossRef]

35. Dinter, A.; Oberwalder, C.; Kabouw, P.; Coulson, M.; Ernst, G.; Leicher, T.; Miles, M.; Weyman, G.; Klein, O. Occurrence and distribution of earthworms in agricultural landscapes across Europe with regard to testing for responses to plant protection products. *J. Soil Sedim.* **2013**, *13*, 278–293. [CrossRef]

36. Geraskina, A.P. Restoration of earthworms community (*Oligochaeta: Lumbricidae*) at sand quarries (Smolensk Oblast, Russia). *Ecol. Quest.* **2019**, *30*, 1–13. [CrossRef]

37. Tripathi, G.; Kachhwaha, N.; Dabi, I.; Singh, J. Earthworms as bioengineers. In *Frontiers in Ecology Research*; Nova Publishers: New York, NY, USA, 2007; p. 227.

38. Iordache, M.; Borza, I. Relation between chemical indices of soil and earthworm abundance under chemical fertilization. *Plant Soil Environ.* **2010**, *56*, 401–407. [CrossRef]

39. Chauhan, R.P.; Pande, K.R.; Shah, S.C.; Dhakal, D.D. Soil properties and earthworm dynamics affected by land use systems in western Chitwan, Nepal. *J. Inst. Agric. Anim. Sci.* **2015**, *33–34*, 123–128. [CrossRef]

40. Khan, M.U.; Ahmed, M.; Nazim, K. The population behavior of earth worm (*Oheritema posthuma* Kinberg) under the influence of industrial waste. *FUUAST J. Biol.* **2017**, *7*, 1–8.

41. Jayathilakan, K.; Sultana, K.; Radhakrishna, K.; Bawa, A.S. Utilization of byproducts and waste materials from meat, poultry and fish processing industries: A review. *J. Food Sci. Tech.* **2012**, *49*, 278–293. [CrossRef] [PubMed]

42. Yunta, F.; Di Foggia, M.; Bellido-Díaz, V.; Morales-Calderón, M.; Tessarin, P.; López-Rayo, S.; Tinti, A.; Kovács, K.; Klencsár, Z.; Fodor, F.; et al. Blood meal-based compound. Good choice as iron fertilizer for organic farming. *J. Agric. Food Chem.* **2013**, *61*, 3995–4003. [CrossRef] [PubMed]

43. Kowalski, Z.; Makara, A.; Banach, M. Krew zwierzęca, metody jej przetwarzania i zastosowanie. *Tech. Trans. Chem.* **2011**, *108*, 87–105.

44. Neuhauser, E.F.; Loehr, R.C.; Milligan, D.L.; Malecki, M.R. Toxicity of metals to the earthworm *Eisenia fetida*. *Biol. Fert. Soil* **1985**, *1*, 149–152. [CrossRef]

45. Spurgeon, D.J.; Hopkin, S.P. The effects of metal contamination on earthworm populations around a smelting works: Quantifying species effects. *Appl. Soil Ecol.* **1996**, *4*, 147–160. [CrossRef]

46. Zhao, L.; Qiu, J.P. Aluminum bioaccumulation in the earthworm and acute toxicity to the earthworm. In Proceedings of the 4th International Conference on Bioinformatics and Biomedical Engineering, Chengdu, China, 18–20 June 2010.

47. Stoven, K.; Schnug, E. Long term effects of heavy metal enriched sewage sludge disposal in agriculture on soil biota. *Landbauforsch. Volkenrode* **2009**, *59*, 131–138.

48. Rastetter, N.; Rothhaupt, K.O.; Gerhardt, A. Ecotoxicological Assessment of Phosphate Recyclates from Sewage Sludges. *Water Air Soil Pollut.* **2017**, *228*, 171. [CrossRef]

49. Elyamine, A.M.; Afzal, J.; Rana, M.S.; Imran, M.; Cai, M.; Hu, C. Phenanthrene mitigates cadmium toxicity in earthworms *Eisenia fetida* (epigeic specie) and *Aporrectodea caliginosa* (endogeic specie) in soil. *Int. J. Environ. Res. Public Health* **2018**, *15*, 2384. [CrossRef] [PubMed]

50. Yahyaabadi, M.; Hamidian, A.H.; Ashrafi, S. Dynamics of earthworm species at different depths of orchard soil receiving organic or chemical fertilizer amendments. *Eurasian J. Soil Sci.* **2018**, *7*, 318–325. [CrossRef]

51. Jastrzębska, M.; Kostrzewska, M.K.; Makowski, P.; Treder, K.; Marks, M. Effects of ash and bone phosphorus biofertilizers on *Bacillus megaterium* counts and select biological and physical soil properties. *Pol. J. Environ. Stud.* **2015**, *24*, 1603–1609. [CrossRef]

52. Jastrzębska, M.; Kostrzewska, M.K.; Makowski, P.; Treder, K.; Jastrzębski, W.P. Functional properties of granulated ash and bone-based phosphorus biofertilizers in the field assessment. Part 3. Impact on selected properties of soil environment of winter wheat. *Przem. Chem.* **2016**, *95*, 1591–1594.

53. Jastrzębska, M.; Kostrzewska, M.K.; Makowski, P.; Treder, K.; Jastrzębski, W.P. Granulated phosphorus fertilizer made of ash from biomass combustion and bones with addition of *Bacillus megaterium* in the field assessment. Part 3. Impact on selected properties of soil environment of winter wheat. *Przem. Chem.* **2017**, *96*, 2180–2183.

 *agronomy*

Article

A Comparative Study of Field Nematode Communities over a Decade of Cotton Production in Australia

Oliver Knox [1,*], David Backhouse [1] and Vadakattu Gupta [2]

[1] School of Environmental and Rural Sciences, University of New England, Armidale, NSW 2350, Australia; dbackhou@une.edu.au

[2] Agriculture & Food, CSIRO, Adelaide, SA 5064, Australia; Gupta.Vadakattu@csiro.au

* Correspondence: oknox@une.edu.au; Tel.: +61-2-6773-2946

Received: 23 December 2019; Accepted: 13 January 2020; Published: 15 January 2020

Abstract: Soil nematode populations have the potential to indicate ecosystem disturbances. In response to questions about nematode interactions with soilborne diseases and whether genetically modified cotton altered nematode populations, several fields in the Namoi cotton growing area of Australia were sampled between 2005 and 2007. No significant interactions were observed, but nematodes numbers were low and postulated to be due to the use of the nematicide aldicarb. Aldicarb was removed from the system in 2011 and in 2015 funding allowed some fields to be resampled to determine if there had been a change in the nematode numbers following aldicarb removal. No significant changes in the total nematode numbers were observed, implying that the removal of aldicarb had little impact on the total nematode population size. However, an increase in plant parasitic nematodes was observed in both fields, but the species identified and the levels of change were not considered a threat to cotton production nor driven solely by altered pesticide chemistry. Additionally, greater numbers of higher order coloniser-persisters in the 2015 samples suggests that the current cotton production system is less disruptive to the soil ecosystem than that of a decade ago.

Keywords: axonchium; helicotylenchus; tylenchorhynchus; pratylenchus; reniform; vertosol; gossypium

---

1. Introduction

The use of pesticides often courts controversy and remains an issue that often results in political intervention [1,2]. Changes in the regulatory processes of both the EU and the United States EPA brought about a decision from Bayer to halt production of aldicarb, a nematicide developed in the 1970s, by 2014 and for complete removal of the product by 2018 [3]. Aldicarb was utilised on a range of crops, but primarily in Australia in cotton, sugar cane and citrus [4,5].

Australian cotton systems have historically been without the nematode related production issues experienced by other cotton producing nations [6], although the presence of the reniform nematode, *Rotylenchus reniformus* [7], in the Theodore production area of Queensland highlights that this status can change. As a consequence of this, aldicarb was not registered for nematode control, but for early season control of aphids, mirids, jassids, mites, wireworms and thrips that aldicarb's systemic activity offered whilst retaining beneficial populations [8]. Control of these early season pests following the removal of aldicarb from Australia in 2011 has been provided either through the optional use of neonicotinoids, in the form of Cruiser® (active ingredient (a.i.) thiomethoxam, Syngenta) [9], or through the continued or adopted use of the organophosphates and carbamates, such as phorate and carbosulfan, respectively. The impact of neonicotinoids on entomopathogenic nematodes has been reported to have limited impact on reproduction [10,11], which might imply limited effects on other free living soil nematodes [10,11]. The organophosphate and carbamates are known to have nematicidal activity particularly against reniform, lesion and root-knot nematodes [12,13], but existing work has been on sandy soils, not in clay vertosols. Additionally, impacts beyond the targeted pest nematode population have either not been undertaken [12,14] or found no difference [13].

Adoption of the synthetic pyrethroids to control of wireworm and mirids offers protection to above and below ground herbivorous damage, however, their impact on nematodes is negligible [5,15]. This assumption is based on the facts that no deleterious effects from synthetic pyrethroids have been found on entomophathogenic nematodes [16–18]. However, when pyrethroids were introduced to aquatic systems nematodes flourished [19], although *Daptonema trabeculosum* was found to be sensitive to permethrin [15].

In the USA, aldicarb has been replaced in the cotton production system with either Avicta® seed treatments (a.i. abamectin, thiamethoxam, mefenoxam and fludioxanil, Syngenta) in possible conjunction with Velum® (a.i. fluopyram and imadicloprid, Bayer CropScience) or the use of Vydate® (a.i. anticholinesterase, DuPont). At present, these products are not licensed for Australian cotton where rotations and management conditions to promote rapid cotton establishment are the predominant forms of nematode control [20,21].

In our initial nematode work in the Namoi in between 2005 and 2007, the low numbers of recovered nematodes (<5 nematodes/g soil) were hypothesised as being due to the systemic use of aldicarb [22,23]. This assumption was based on the impact aldicarb has on free living nematodes in culture and under carrots [11,24]. However, despite being initially developed as a nematicide, aldicarb has been rarely studied, in relation to free living nematodes [11], does not affect free living nematodes under potato [25] and we could find no published evidence of its impact under cotton rotations. With changes in funding, movement of staff and the removal of aldicarb in 2011, we were unable to test our hypothesis directly, instead resampling fields in in the upper and lower Namoi valley in 2015, which were originally sampled in 2005 and 2007 and for which nematode community analysis had been undertaken [22,26]. The nematode communities were assessed and compared between the sampling years to determine if the nematode numbers had increased with the removal of aldicarb and if there had been changes in the nematode population structure. The results are discussed within the context of the potential for effects on the Australian cotton production system and the ecological significance of the observations.

2. Materials and Methods

2.1. Soil Characteristics and Nematode Sampling

Field A: In July 2005 and June of 2007, a field in the lower Namoi (field A) was sampled as part of investigations into non-target effects of genetically modified (GM) cotton on soil microbiology. The field soil is a grey vertosol, 52% clay, pH 8.2 and 200 m above sea level. The mean annual maximal temperatures is 26 °C and minimum 12 °C and the area receives 660 mm of summer dominant rainfall. In the field, samples were collected from under each variety being cultivated, resulting in 16 samples in 2005 and 12 in 2007, with sites evenly spaced along 180 m of the plant line. Approximately one kilogram of topsoil was taken to a depth of 15 cm at each site from under mature cotton. In March, 2015, this field was resampled when it was again under cotton, using field maps of the 2007 trial to return to approximately the same location except that only six samples were taken from the plant line at equidistant points from the tail to head ditch with the field having been planted under only one variety. This field had been in a cotton–wheat rotation, with cotton planted in October of every even year. Aldicarb had been applied at cotton sowing at a standard rate of 7 kg Temik®/ha (1.05 kg a.i.) for thrips control with the final application made in October of 2010. In 2012 and 2014, phorate was applied with cotton sowing as 6 kg Thimet®/ha (600 g/ha a.i.). Neither chemical was used in the wheat phase of the rotation.

Field B: In late October of 2005, soil was sampled from a field in the upper Namoi (field B) as part of an investigation into nematode interactions with verticillium wilt. This field is a black vertosol, 65% clay, pH 8.5 and 270 m above sea level. Mean maximum and minimum temperatures are 12 and 27 °C, respectively, with the area receiving roughly 640 mm of summer dominant rain. One kilogram of surface soil to a depth of 15 cm was recovered from the plant line of cotton seedlings. Briefly, sample points were established from both the Northern and North-Western corners of the field by walking a 20 m by 10 row transect into the crop and taking a sample. The transect walk was then repeated until six samples had been gathered from each entry point. In March, when the field was under mature cotton and again in June of 2015 after picking and root cutting, we collected samples close to the original sampling points, based on field notes and discussions with the farmer. This field had predominantly been under a cotton–cotton–wheat rotation since 1988, although sorghum had been introduced in place of wheat in 2009, 2013 and 2014. Aldicarb had been applied as Temik® at 7 kg/ha in every year that cotton was sown, resulting in aldicarb application in 13 out of 28 years, with the last application in 2011.

Cultivations varied between fields due to differences in the rotations, but both had been subjected to pupae busting, a minimal cultivation to a depth of 10 cm at least 30 cm either side of the plant line, post cotton crop harvesting and had been subjected to bed reformation in the spring prior to cotton planting.

2.2. Soil Analysis

In all cases, field sampled soil was placed in plastic bags and returned in a chilled ice box to the laboratory. In the laboratory, the samples were sieved through a 2 mm sieve and a 300 g subsample was sent within 48 h of samples being taken in the field to Biological Crop Protection (Moggill, Queensland, Australia) for nematode community analysis. Briefly, the soil moisture content was determined gravimetrically and 200 mL of soil was weighed and used to establish Whitehead trays for nematode extraction. Nematodes were subsequently recovered from the water solution within the trays and assessed to determine nematode abundance. A sample of approximately 120 nematodes from the count were identified to genus and, in the case of the plant parasitic nematodes, to species where possible to facilitate community compositional analysis [27]. Recovered nematode data were analysed both as recovered numbers and as the number of nematodes present per gram of dry weight equivalent of soil to mitigate moisture and soil porosity differences.

2.3. Root Tissue Analysis

Roots were collected from all samples during the sieving process and the root tissue was cleared using the NaOCl and acid fuchsin method of Byrd et al. [26,28]. Roots were spread over a 1 cm gridded Petri dish and examined under a stereo microscope (20 to 45 x magnification) for the presence of nematodes.

2.4. Community Comparisons and Statistical Analysis

The nematode community data from the 2005, 2007 and 2015 field samples were tabulated. Comparative analyses for the free living nematodes and between the plant parasitic nematode types were conducted on either raw or percentage compositional data, respectively, with multiple Wilcoxon rank-sum tests between all possible pairwise comparisons. Significance in differences of the median values was taken at the level of $p < 0.05/x$, where x represented the number of groups within any series of pairwise comparisons. This decision was based on the existence of small sample sets for each field and a lack of normality of the data. The nematode channel ratio (NCR) [29] was calculated from the bacterial and fungal trophic group composition of the samples. Additional community composition and change was assessed using the Nematode INdicator Joint Analysis (NINJA) web based program [30] with probability of similarity of mean outcomes assessed with ANOVA, with significance taken at $p < 0.05$. This on-line tool was also used to generate maturity index (MI), Plant Parasitic Index (PPI), enrichment (EI) and structural indexes (SI) for the samples [31,32].

3. Results

3.1. Soil Sample and Total Nematode Comparisons

The 200 mL soil samples had an averaged dry weight equivalent of 126.5 g (stdev = 4.5, $n = 30$) for field A and 134.5 g (stdev = 8.8, $n = 18$) for field B over the period of assessment with no apparent statistical difference between weights with sampling time or field, however, moisture content varied between 24% and 35%. The total number of nematodes recovered per 200 mL of soil ranged from 267 to 2944, with an average of 1194, mode of 371 and standard deviation of 609 and standard error of 85. Analysis of the total recovered nematodes did not indicate any significant difference in nematodes/g assessed either within fields, between years or in combination (Table 1), but were detected for many nematode ecological indexes and footprints (Table 2), primarily due to changes in the nematode population structure recorded in 2015 in field B.

Table 1. Mean nematode counts of total free living nematodes, per g dry weight equivalent of soil and the percentage of plant parasitic from 200 mL soil Whitehead tray recoveries of samples collected in cotton fields A and B in the Namoi valley. The percentage contributions of the stunt (*Merlinius* and *Tylenchorhynchus* spp.), lesion (*Pratylenchus* sp.) spiral (*Helicotylenchus* sp.) and dagger nematodes to the plant parasitic nematodes within samples and years are given. Similarities in the plant parasitic population are assessed with Wilcoxon rank-sum tests and significantly similar medians are indicated with the same upper case letter.

Field	Year	Nematodes Per 200 mL	Per g/Soil	% Plant Parasitic	% Stunt	% Lesion	% Spiral	% Dagger
A	2005	1064	8.5	2.9	92.4 [A]	1.3 [A]	nd	nd
	2007	791	6.3	1.0	41.8 [B]	49.9 [B]	nd	nd
	2015	1319	10.2	1.6	11.3 [B]	81 [B]	7.7	nd
		ns	ns	ns	$p < 0.01$	$p < 0.001$	ns	ns
B	2005	1229	9.8	1.1	100.0	nd	nd	nd
	2015	1687	11.8	7.7	61.7	nd	36.8	1.4
		ns	ns	$p < 0.001$	$p < 0.001$		$p < 0.01$*	ns

* statistical analysis in cases where the nematode was previously not detected assumes a 0 value in the samples of those years. No detection within the samples is indicated by 'nd' and 'ns' indicates no significant difference.

Table 2. Summary mean, standard deviations (SD) and corresponding ANOVA p values from the Nematode INdicator Joint Analysis (NINJA) of the field analysed samples from 2005, 2007 and 2015 in field A and 2005 and 2015 in field B.

Index Name		Field A 2005	Field A 2007	Field A 2015	Field B 2005	Field B 2015	ANOVA p Value
Maturity Index	mean	2.3	2.2	2.2	2.1	2.4	<0.001
	SD	0.2	0.2	0.1	0.1	0.1	
Plant Parasitic Index	mean	2.4	2.6	2.1	2.3	3.2	<0.001
	SD	0.3	0.3	0.1	0.4	0.5	
Enrichment Index	mean	35.6	26.1	37.4	44.3	49.2	0.001
	SD	6.2	17.6	14.3	10.6	12.9	
Structure Index	mean	51.3	38.5	47.5	35.5	68.4	<0.001
	SD	15.2	13.6	7.6	8.8	11.7	
Nematode channel ratio	mean	0.7	0.9	0.9	0.7	0.9	<0.001
	SD	0.1	0.1	0.0	0.1	0.1	

Table 2. *Cont.*

Index Name		Field A 2005	Field A 2007	Field A 2015	Field B 2005	Field B 2015	ANOVA *p* Value
Herbivore footprint	mean	1.5	1.5	3.2	2.0	21.0	<0.001
	SD	0.6	0.6	1.4	0.8	10.5	
Fungivore footprint	mean	3.2	1.3	0.7	2.2	0.8	<0.001
	SD	1.5	0.7	0.5	0.8	0.5	
Bacterivore footprint	mean	14.4	22.5	24.2	24.5	15.2	0.009
	SD	3.8	13.4	5.3	10.6	3.4	
Predator footprint	mean	1.6	0.5	0.7	0.0	0.7	0.09
	SD	1.9	1.0	0.9	0.0	0.9	
Omnivore footprint	mean	11.8	7.7	5.3	6.6	5.7	0.044
	SD	8.6	3.8	2.4	5.8	2.8	
Total number	mean	132.6	118.8	120.3	119.2	123.8	0.045
(nematode/200 mL)	SD	11.6	15.7	9.0	15.0	11.4	

3.2. Plant Parasitic Nematode Populations

The percentage of the nematode population representing plant parasitic nematodes had not changed in field A and was reflected in the PPI scores for the field, which averaged 2.38, 2.56 and 2.09 for 2005, 2007 and 2015, respectively. However, the PPI had significantly ($p < 0.001$) increased in field B from 2.29 in 2005 to 3.18 in 2015. Additionally, the composition of plant parasitic nematodes, in terms of the abundance of specific parasitic genera, revealed changes in both fields. For example, in the field B there was and remained no evidence of lesion nematodes (*Pratylenchus* sp.), but a significant decrease in stunt (*Merlinius* and *Tylenchorhynchus* spp.) and an increase in spiral (*Helicotylenchus* sp.) nematodes was observed. In field A, spiral nematodes were not observed in 2005 and 2007 samples, but were found in the 2015 samples at >0.2% of the total nematode population. Stunt nematodes were significantly ($p < 0.001$) higher in both fields in 2005 than in other sampling years, whilst the proportion of lesion nematodes increased with time in field A (Table 1). Data on the abundances of the ectoparasites, semi-endoparasites and migratory endoparasites as their % composition of the herbivore assemblage implied that within field A the migratory endoparasites increased as the ectoparasites were reduced, whilst in field B the semi-endoparasties appeared to have replaced the migratory endoparasites (Figure 1).

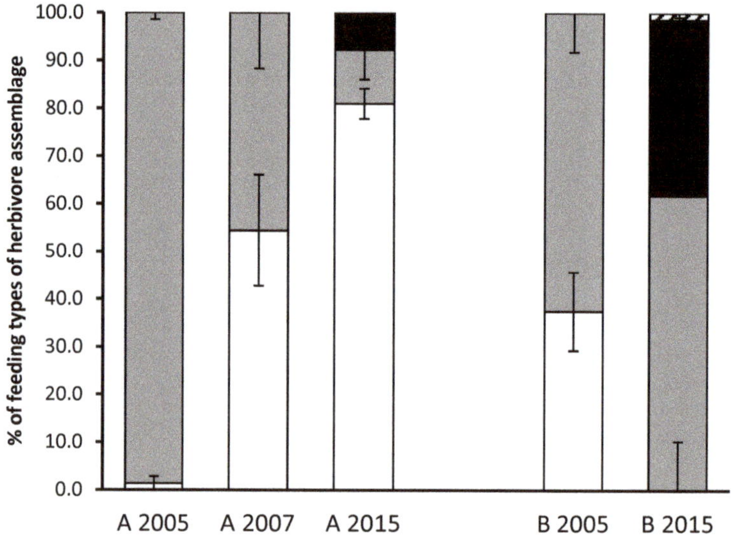

Figure 1. The percentage of the migratory endoparasites (e.g., *Pratylenchus*, white), ectoparasitic (e.g., *Tylenchorhynchus*, grey), semi-endoparasitic (e.g., *Helicotylenchus*, black) and ectoparasitic (e.g., *Xiphenema*, dashed) feeding types of the herbivorous nematodes assemblage identified from 200 mL soil samples from field A in 2005, 2007 and 2015 and field B in 2005 and 2015. Number of samples and time of year differed between years with error bars representing the standard error of the means.

3.3. Nematode Community Assemblages

Community analysis with NINJA indicated that there was significant ($p < 0.05$, ANOVA) difference in the maturity, plant parasitic, enrichment and structural indexes and the herbivore, fungivore, bacterivore and omnivore footprints within the assessed field material (Table 2). The changes in the assessed community reflected these differences in terms of shifts in the relative proportions of omnivore, predatory, bacterivores, fungivores and herbivorous nematodes present (Figure 2) as well as in changes to the composition of the herbivorous nematode assemblage (Figure 1). Whilst changes in the structural and enrichment status of the samples were both significant (Table 2), graphical representation of the data (Figure 3) supported an improvement in maturity of the analysed ecosystem rather than nutrient enrichment, due to an increase in the number of higher order coloniser-persisters in the samples. This was particularly evident for field B between 2005 and 2015 (Figure 3). NCR analysis indicated similar scores between fields, but that the 2005 samples had a lower ratio than the populations of subsequent samples in both fields (Table 2).

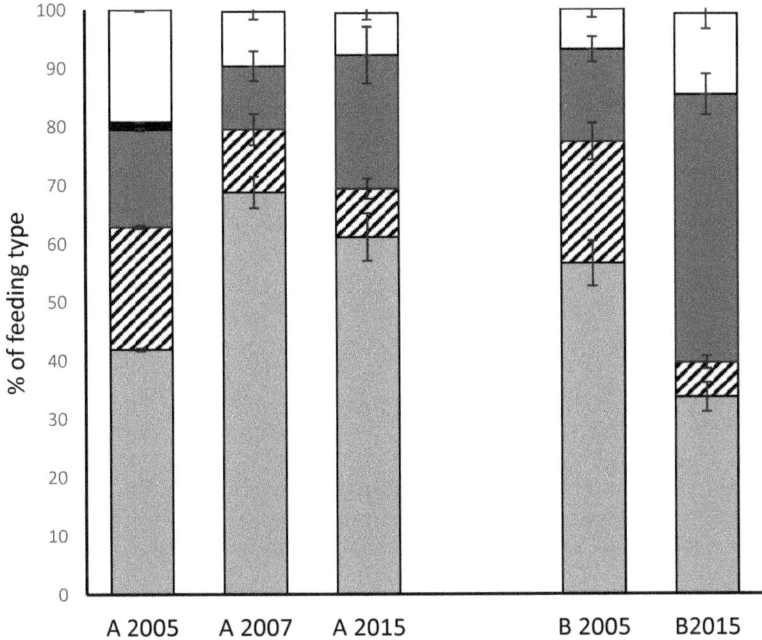

Figure 2. Average percentage of the total recovered omnivorous (white), predatory (black), bacterivorous (light grey), fungivorous (stripped) and herbivorous (dark grey) nematode feeding types as identified from the evaluation of ~120 nematodes from each sample ($n \geq 6$) from field A and field B over each year of sampling. Error bars represent the standard errors of the means.

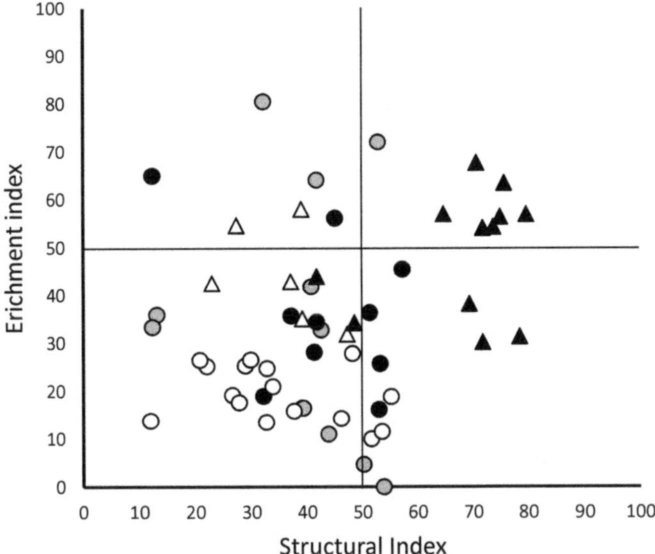

Figure 3. Food web analysis of nematode community assemblages from field A, sampled in 2005 (○), 2007 (◔) and 2015 (●) and field B sampled in 2005 (△) and 2015 (▲). The Enrichment index parallels with the nutrient enrichment whilst the structural index correlates with the maturity of the ecosystem.

3.4. Root Tissue Observations

No nematodes were observed within the cleared and stained root tissue from the 2015 samples, which was in contrast to the observation and recovery of *H. dihystera* from roots in field B and of Rhabditea and Aphelenchidea from roots in field A in 2005.

4. Discussion

In general, abundances of total nematodes in soil supporting Australian cotton systems, as observed in 2005 and 2007 [26,33], are considered low [34]. In addition to this, Australian cotton production systems have not reported nematode issues, with the exception of the recent and localized occurrence of the reniform nematode [7], and this was partly attributed to the widespread use of aldicarb in cotton [22,23]. Aldicarb has a highly variable half-life in soil that ranges from a few to 408 days, with more rapid detoxification occurring in anaerobic soils [4,5]. In Australian cotton soils, the half-life is thought to be about a week in surface soils, due to high soil temperatures and the repeated fluctuation between aerobic and anaerobic soil conditions from flood irrigations [4]. With aldicarb absent from these cotton fields for several years, residual compound and active metabolites from historic applications should have fallen below effective levels [35]. In an attempt to discern if this hypothesis was correct, two fields, roughly 160 km apart, in the Namoi valley, which had nematode community data from 2005 and 2007, were reassessed in 2015.

Although observations from the 2015 sampling indicated that significant changes in the composition of nematode communities were occurring (Figure 3), the total numbers of nematodes supported within the vertosols had not changed (Table 1). This was taken as indication that aldicarb had not imposed a limitation on the population size as initially hypothesised, which is in keeping with other work where pesticide changes had not altered nematode population size, but had been associated with a change in species richness [36,37]. Whilst the implications of other variations in the assessed fields' management systems, such as differences in clay content, irrigation strategies, rotational histories and periods of fallow, could not be investigated from the field records available, it was noted that between the two fields the frequency of fallows occurring post wheat and prior to the return to cotton in the rotation varied [38]. Periods of long fallow of over 7 months in Australian grains production systems, which can incorporate cotton, have been previously reported as causing a reduction in the free living nematode population and altering the nematode channel ratio [29] in favour of a fungal dominated decompositional community [39]. However, the populations analysed in these fields indicated a move to more bacterially dominated decompositional communities over time (Table 2).

Whilst we saw little change in the total free living nematode population across our samples, changes in the nematode community composition were noted in the herbivorous assembly in field B whilst herbivorous nematodes remained unchanged in field A. In a study in Slovakian, maize fields increasing insecticidal chemistry to five times the recommended dose did not significantly alter the nematode communities, but season of assessment did [37]; however, this trial did not interrogate other management decisions. When comparing results from these cotton fields to existing studies [37,39,40], it becomes apparent that there is a requirement for further systematic interrogation of the production systems in order to identify the drivers of nematode community change.

Changes in cotton production practices are also possible causes of the observed differences in the nematode communities within these fields over the last decade [41]. Since 2005, both farms have experienced drought that has seen both differences in the amount and quality of the water used for cotton irrigation in different years, which could have influenced nematode communities [42]. There has also been a change in the preferred cultivar material from cultivars based on the Sicot 189 family in 2005/7 to those of Sicot 74 and 75 in 2015 along with changes in pesticide use and nutrition management [41], which has included the loss of aldicarb from the Australian cotton production system. Additionally, sampling was not possible around the time of aldicarb removal from the system due to funding, staffing movements and that sampling across the two fields occurred at different times within the cotton phase of the rotation, due to weather constraints that were unavoidable. These issues further highlight that gaps exist in our knowledge of nematodes within Australian vertosols over temporal periods.

Knowledge about the long-term changes in nematode communities due to changing crop management practices would help in the development of options to avoid unexpected threats in addition to providing insights into the ecology of soil fauna in production systems with multiple crop, chemical and physical factors potentially influencing abundance and composition [43]. So whilst the main drivers of nematode community change remain elusive, the nature of the differences between fields and study periods highlighted the continued need for vigilance and the imposition of the 'come clean, go clean' farm hygiene strategy, as currently promoted throughout the Australian cotton industry. This strategy is required to continue to limit the spread of potential problem nematodes, such as the reniform nematode, which is causing cotton production issues in Theodore [7], but remains undetected in New South Wales (NSW). However, the presence of *H. dihystera* within field A and *Xiphenema* sp. in field B in the 2015 samples was noted as neither had been previously detected there. Whilst it is possible that these nematodes were not previously observed due to scarcity, the possibility that they were introduced through soil movement on contaminated machinery over the intervening decade remains plausible.

Changes in other members of the herbivorous nematode population were also noted. *T. ewingi*, was still isolated from both fields, but in field B *T. ewingi* was significantly reduced as a percentage of the plant parasitic population due to an increase in soil recovery of *H. dihystera* (Table 1). This change was hypothesised as being due to rotational differences, which included the incorporation of sorghum into the rotation of field B. This hypothesis was based on both *Tylenchorhynchus* and *Helicotylenchus* spp. being known to survive on wheat [43] and having both been recorded on wheat and sorghum in Australia [44]. Additionally, in a >20 year experiment involving continuous sorghum there was little impact on *Tylenchorhynchus* spp., but incorporation of sorghum straw resulted in a significant increase in the number of *Helicotylenchus* spp. recovered [45], which mirrored the observed change in field B.

The isolation of *H. dihystera* within field B was also noted to have changed over the decade. *H. dihystera* was first observed in Australian cotton roots collected from field B [26], but was absent from the soil samples in 2005. However, these observations were reversed in 2015 with *H. dihystera* only observed in soil. This observation could possibly be linked to the difference in the time of sampling [37] and a reduction in the number of samples, but might also be a function of the maturity of the cotton roots. More likely though is *H. dihystera* ability to feed on sorghum as either an endo or ectoparasite [45,46] and that sorghum was planted into the field B rotation in three of the previous five years to the 2015 sampling.

The other plant parasitic nematode shift considered to be of note was that of the lesion nematode, mostly *P. thornei*, which remained absent in field B, but had significantly increased in numbers in field A. Although still not considered an issue for cotton production in Australia, establishment of a population of around the levels found in 2015 without appropriate management could become an issue for grain crops grown in rotation with cotton [47,48].

Out with the changes in the plant parasitic populations, there was an increase in general maturity index of the community in the 2015 soil samples, suggesting an increase in the abundance of higher order coloniser-persister (C-P) nematodes. This change was particularly evident with the increase in the numbers of *Axonchium* sp., although it was echoed to a lesser extent in other nematodes with C-P scores of >3 [31]. The *Axonchium* nematodes increased from 0.14% to 0.28% of the population in field A, but in field B they increased from 0.42% to 17% of the total population and in some samples represented 40% of the total free living nematode population. Members of the genus *Axonchium* pose something of enigma, because the lack of a clearly identifiable mouth part makes them hard to assign to a specific trophic group. This has seen *Axonchium* associated with either bacterial, root hair and therefore plant parasitic or predatory feeding patterns [49,50]. Given the increase in these nematodes in field B, it would be prudent to establish the exact feeding strategy of these nematodes, as changes in assignation of feeding strategy to a fungivore or omnivore, rather than an herbivorous ectoparasite, increased the maturity and structural index, whilst reducing the plant parasitic index for field B. However, altering the assigned feeding type for *Axonchium* had little to no impact on either the channel or enrichment index and no effect on field A analysis, where they were less abundant in the samples.

From a production stand point, the apparent rise in plant associated and parasitic nematodes could be seen as grounds for concern, especially in the absence of any chemical or cultivar control options, but at the same time the increase in the maturity index of the populations (Figure 3), partly though changes in predatory nematodes, could be indication of more persistent and stable populations that might self-regulate any potential production threat [37,42]. Although most of the samples still exemplify a state of degradation, based on the quadrat in which they occur [32], there does appear to be a trend toward a trajectory in both enrichment and structural indexes (Figure 3). This observation implies that between 2005 and 2015 the examined cotton production systems are moving toward more opportunistic bacterial feeding strategies, based on the enrichment index, whilst the improvement in the structural index implies a less disturbed soil food web and improved trophic interactions [51]. However, nematodes of the higher order trophic groups, which drive these developments, are known to be easily disrupted by soil cultivation [51], making this a potentially unreliable control mechanism under existing cotton production strategies that still involve some form of tillage.

In general, these observations indicate a continuing change in the nematode populations in the Australian cotton fields sampled, probably due to changes in soil management, rotational variation and seasonal environmental conditions [37,41,45], whereas the impact from pesticides is perhaps not as important as originally hypothesised [22]. However, the scale of the current assessment highlights a need for more intensive sampling and for an improved understanding of the genera present. Whilst changes in the herbivorous nematode populations in these NSW fields implies limited current threat to cotton production in these areas, the risk of movement of the reniform nematode from Queensland and the absence of available nematicidal chemistry would caution that continued monitoring and vigilance is warranted.

Author Contributions: O.K. undertook the field sampling, analysis and manuscript preparation. D.B. provided project delivery assistance, technical and editorial support. V.G. assisted with the 2005 and 2007 sample analysis, strategy for the 2015 analysis and manuscript editorial support. All authors have read and agreed to the published version of the manuscript.

Funding: This work was undertaken as part of the activities of the Cotton Hub at UNE with funding provided by the University of New England and the Cotton Research and Development Corporation under UNE1403 and UNE2001. The Initial surveys were conducted with funding from the Cotton Catchment and Communities CRC and CRDC with assistance from staff at CSIRO and NSW DPI. Nematode extraction and analysis was conducted by Biological Crop Protection, Moggill, Queensland for all samples other than root material.

Conflicts of Interest: The authors declare no conflict of interest.

References

1. Skevas, T.; Lansink, A.; Stefanou, S.E. Designing the Emerging Eu Pesticide Policy: A Literature Review. *NJAS-Wagening. J. Life Sci.* **2013**, *64*, 95–103. [CrossRef]
2. Bielza, P.; Denholm, I.; Sterk, G.; Leadbeater, A.; Leonard, P.; Jørgensen, L.N. Declaration of Ljubljana #8211; the Impact of a Declining European Pesticide Portfolio on Resistance Management. *Outlooks Pest Manag.* **2008**, *19*, 246–248.
3. Erickson, B. Pesticides: Bayer Cropscience, Epa Agree to Phase out Use of Aldicarb. *Chem. Eng. News* **2010**, *88*. [CrossRef]
4. Pesticides, A.; Authority, V.M. The Nra Review of Aldicarb. In *Existing Chemical Review Program*; National Registration Authority for Agricultural and Veterinary Chemicals: Canberra, Australia, 2001.
5. Cox, C. Aldicarb. *J. Pestic. Reform* **1992**, *12*, 31–35.
6. Robinson, A.F. Nematode Management in Cotton. In *Integrated Management and Biocontrol of Vegetable and Grain Crops Nematodes*; Ciancio, A., Mukerji, K.G., Eds.; Springer: Dordrecht, The Netherlands, 2007; pp. 149–182.

7. Bauer, B.; Smith, L.; Scheikowski, L.; Lehane, J.; Cobon, J.; O'Neill, W. Reniform Nematode Surveys in Central Queensland Cotton. In Proceedings of the 17th Australian Cotton Conference, Gold Coast, Australian, 5 August 2014.

8. Farrell, T. *Cotton Pest Management Guide 2007–2008*; NSW Department of Primary Industries: Orange, Australian, 2007.

9. CSD. New Cotton Seed Treatment Reduces in-Furrow Woes. *Seeds Thought* **2011**, *10*, 19.

10. Koppenhöfer, M.A.; Fuzy, E.M. Early Timing and New Combinations to Increase the Efficacy of Neonicotinoid–Entomopathogenic Nematode (Rhabditida: Heterorhabditidae) Combinations against White Grubs (Coleoptera: Scarabaeidae). *Pest Manag. Sci.* **2008**, *64*, 725–735. [CrossRef] [PubMed]

11. Tahseen, Q.; Shamim Jairajpuri, M.; Ahmad, I. Nematicidal Impact on the Reproductive Biology of *Mesorhabditis Cranganorensis. Afro-Asian J. Nematol.* **1996**, *6*, 184–187.

12. Meher, C.H.; Gajbhiye, V.T.; Singh, G.; Kamra, A.; Chawla, G. Persistence and Nematicidal Efficacy of Carbosulfan, Cadusafos, Phorate, and Triazophos in Soil and Uptake by Chickpea and Tomato Crops under Tropical Conditions. *J. Agric. Food Chem.* **2010**, *58*, 1815–1822. [CrossRef]

13. Wada, S.; Toyota, K.; Takada, A. Effects of the Nematicide Imicyafos on Soil Nematode Community Structure and Damage to Radish Caused by Pratylenchus Penetrans. *J. Nematol.* **2011**, *43*, 1–6.

14. Khan, R.M.; Zaidi, B.; Haque, Z. Nematicides Control Rice Root-Knot, Caused by Meloidogyne Graminicola. *Phytopathol. Mediterr.* **2012**, *2012*, 298–306.

15. Soltani, A.; Louati, H.; Hanachi, A.; Salem, F.B.; Essid, N.; Aissa, P.; Mahmoudi, E.; Beyrem, H. Impacts of Permethrin Contamination on Nematode Density and Diversity: A Microcosm Study on Benthic Meiofauna from a Mediterranean Coastal Lagoon. *Biologia* **2012**, *67*, 377–383. [CrossRef]

16. Hara, A.H.; Kaya, H.K. Effects of Selected Insecticides and Nematicides on the in Vitro Development of the Entomogenous Nematode Neoaplectana Carpocapsae. *J. Nematol.* **1982**, *14*, 486–491. [PubMed]

17. Sabino, P.H.; Sales, F.S.; Guevara, E.J.; Moino, A., Jr.; Filgueiras, C.C. Compatibility of Entomopathogenic Nematodes (Nematoda: Rhabditida) with Insecticides Used in the Tomato Crop. *Nematoda* **2014**, *1*. [CrossRef]

18. Zhang, L.; Shono, T.; Yamanaka, S.; Tanabe, H. Effects of Insecticides on the Entomopathogenic Nematode Steinernema Carpocapsae Weiser. *Appl. Entomol. Zool.* **1994**, *29*, 539–547. [CrossRef]

19. Cox, C. Cyfluthrin. *J. Pestic. Reform* **1994**, *14*, 28–34.

20. Allen, S.; Smith, L.; Scheikowski, L.; Gambley, C.; Sharman, M.; Maas, S.; Kirkby, K.; Lonergan, P. Common Diseases of Cotton. In *Cotton Pest Management Guide 2014–15*; Susan, M., Ed.; Greenmount Press: East Toowoomba, Australia, 2014; pp. 118–121.

21. Blessitt, A.J.; Stetina, S.R.; Wallace, T.P.; Smith, P.T.; Sciumbato, G.L. Cotton (Gossypium Hirsutum) Cultivars Exhibiting Tolerance to the Reniform Nematode (Rotylenchulus Reniformis). *Int. J. Agron.* **2012**. [CrossRef]

22. Knox, G.O.G.; Anderson, C.M.T.; Nehl, D.B.; Gupta, V.V.S.R. Observation of Tylenchorhynchus Ewingi in Association with Cotton Soils in Australia. *Plant Dis.Notes* **2006**, *1*, 47–48. [CrossRef]

23. Knox, O.; Anderson, C.; Vadakattu, G.; Seymour, N. Tiny Worms: Nematodes in Australian Cotton. *Aust. Cottongrow.* **2007**, *28*, 10–13.

24. Lue, P.L.; Lewis, C.C.; Melchor, V.E. The Effect of Aldicarb on Nematode Population and Its Persistence in Carrots, Soil and Hydroponic Solution. *J. Environ. Sci. Health B* **1984**, *19*, 343–354. [CrossRef]

25. Sturz, A.; Kimpinski, J. Effects of Fosthiazate and Aldicarb on Populations of Plant-Growth-Promoting Bacteria, Root-Lesion Nematodes and Bacteria-Feeding Nematodes in the Root Zone of Potatoes. *Plant Pathol.* **1999**, *48*, 26–32. [CrossRef]

26. Knox, G.O.G.; Anderson, C.M.T.; Allen, S.J.; Nehl, D.B. Helicotylenchus Dihystera in Australian Cotton Roots. *Australas. Plant Pathol.* **2006**, *35*, 287–288. [CrossRef]

27. Stirling, R.G.; Lodge, G.M. A Survey of Australian Temperate Pastures in Summer and Winter Rainfall Zones: Soil Nematodes, Chemical, and Biochemical Properties. *Soil Res.* **2005**, *43*, 887–904. [CrossRef]

28. Byrd, W.B.; Kirkpatrick, T.L.; Barker, K.R. An Improved Technique for Clearing and Staining Plant Tissues for Detection of Nematodes. *J. Nematol.* **1983**, *15*, 142–143.

29. Yeates, G.W. Nematodes as Soil Indicators: Functional and Biodiversity Aspects. *Biol. Fertil. Soils* **2003**, *37*, 199–210. [CrossRef]

30. Sieriebriennikov, B.; Ferris, H.; de Goede, R.G.M. Ninja: An Automated Calculation System for Nematode-Based Biological Monitoring. *Eur. J. Soil Biol.* **2014**, *61*, 90–93. [CrossRef]

31. Bongers, T. The Maturity Index: An Ecological Measure of Environmental Disturbance Based on Nematode Species Composition. *Oecologia* **1990**, *83*, 14–19. [CrossRef] [PubMed]

32. Ferris, H.; Bongers, T.; de Goede, R.G.M. A Framework for Soil Food Web Diagnostics: Extension of the Nematode Faunal Analysis Concept. *Appl. Soil Ecol.* **2001**, *18*, 13–29. [CrossRef]

33. Knox, G.O.G.; Gupta, V.V.S.R.; Lardner, R. Field Evaluation of the Effects of Cotton Variety and Gm Status on Rhizosphere Microbial Diversity and Function in Australian Soils. *Soil Res.* **2014**, *52*, 203–215. [CrossRef]

34. Yeates, G.W.; Bongers, T. Nematode Diversity in Agroecosystems. *Agric. Ecosyst. Environ.* **1999**, *74*, 113–135. [CrossRef]

35. Coppedge, R.J.; Bull, D.L.; Ridgway, R.L. Movement and Persistence of Aldicarb in Certain Soils. *Arch. Environ. Contam. Toxicol.* **1977**, *5*, 129–141. [CrossRef]

36. Brmez, M.; Ivezic, M.; Raspudic, E.; Tripar, V.; Balicevic, R. Nematode Communities as Bioindicators of Antropogenic Influence in Agroecosystems. *Cereal Res. Commun.* **2007**, *35*, 297–300. [CrossRef]

37. Čerevková, A.; Miklisová, D.; Cagáň, Ľ. Effects of Experimental Insecticide Applications and Season on Soil Nematode Communities in a Maize Field. *Crop Prot.* **2017**, *92*, 1–15. [CrossRef]

38. Farrell, T.; Hulugalle, N.; Gett, V. Healthier Cotton Soils through High Input Cereal Rotations. In Proceedings of the 14th Australian Cotton Conference, Broadbeach, Australia, 12–14 August 2008.

39. Bell, M.; Seymour, N.; Stirling, G.R.; Stirling, A.M.; van Zwieten, L.; Vancov, T.; Sutton, G.; Moody, P. Impacts of Management on Soil Biota in Vertosols Supporting the Broadacre Grains Industry in Northern Australia. *Aust. J. Soil Res.* **2006**, *44*, 433–451. [CrossRef]

40. McSorley, R. Overview of Organic Amendments for Management of Plant-Parasitic Nematodes, with Case Studies from Florida. *J. Nematol.* **2011**, *43*, 69–81. [PubMed]

41. Braunack, M.V. Cotton Farming Systems in Australia: Factors Contributing to Changed Yield and Fibre Quality. *Crop Pasture Sci.* **2013**, *64*, 834–844. [CrossRef]

42. Yeates, W.G.; Wardle, D.A.; Watson, R.N. Responses of Soil Nematode Populations, Community Structure, Diversity and Temporal Variability to Agricultural Intensification over a Seven-Year Period. *Soil Biol. Biochem.* **1999**, *31*, 1721–1733. [CrossRef]

43. Ferris, V.R.; Bernard, R.L. Crop Rotation Effects on Population Densities of Ectoparasitic Nematodes. *J. Nematol.* **1971**, *3*, 119–122.

44. McLeod, R.; Reay, F.; Smith, J. *Plant Nematodes of Australia Listed by Plant and by Genus: Compiled 1994*; Francoise, R., Smyth, J., Eds.; NSW Agriculture: Orange, Australia, 1994.

45. Villenave, C.; Saj, S.; Pablo, A.; Sall, S.; Djigal, D.; Chotte, J.; Bonzi, M. Influence of Long-Term Organic and Mineral Fertilization on Soil Nematofauna When Growing Sorghum Bicolor in Burkina Faso. *Biol. Fertil. Soils* **2010**, *46*, 659–670. [CrossRef]

46. Zahid, I.M.; Gurr, G.M.; Hodda, M.; Nikandrow, A.; Fulkerson, W.J. Orientation, Reproduction and Effect of Spiral Nematode (Helicotylenchus Dihystera) on Growth of White Clover (Cv. Haifa). *Australas. Plant Pathol.* **2002**, *31*, 55–56. [CrossRef]

47. Taylor, P.S.; Vanstone, V.A.; Ware, A.H.; McKay, A.C.; Szot, D.; Russ, M.H. Measuring Yield Loss in Cereals Caused by Root Lesion Nematodes (*Pratylenchus Neglectus* and *P. Thornei*) with and without Nematicide. *Aust. J. Agric. Res.* **1999**, *50*, 617–627. [CrossRef]

48. Thompson, P.J.; Owen, K.J.; Stirling, G.R.; Bell, M.J. Root-Lesion Nematodes (Pratylenchus Thornei and P. Neglectus): A Review of Recent Progress in Managing a Significant Pest of Grain Crops in Northern Australia. *Australas. Plant Pathol.* **2008**, *37*, 235–242. [CrossRef]

49. Small, R.W. A Review of the Prey of Predatory Soil Nematodes. *Pedobiologia* **1987**, *30*, 179–206.

50. Yeates, W.G.; Bongers, T.; de Goede, R.G.M.; Freckman, D.W.; Georgieva, S.S. Feeding Habits in Soil Nematode Families and Genera—An Outline for Soil Ecologists. *J. Nematol.* **1993**, *25*, 315–331. [PubMed]

51. Ferris, H.; Griffiths, B.S.; Porazinska, D.L.; Powers, T.O.; Wang, K.; Tenuta, M. Reflections on Plant and Soil Nematode Ecology: Past, Present and Future. *J. Nematol.* **2012**, *44*, 115–126. [PubMed]

Communication

Distribution and Restricted Vertical Movement of Nematodes in a Heavy Clay Soil

Oliver Knox [1,*], Katherine Polain [1], Elijha Fortescue [1] and Bryan Griffiths [2]

[1]   School of Environmental and Rural Science, University of New England, Armidale 2351, NSW, Australia; kpolain2@une.edu.au (K.P.); elijhafortescue@gmail.com (E.F.)
[2]   Scotland's Rural College, King's Buildings, Edinburgh EH9 3JG, UK; beinnaghlo@gmail.com
*   Correspondence: oknox@une.edu.au; Tel.: +61-2-67732946

Received: 24 December 2019; Accepted: 1 February 2020; Published: 4 February 2020

Abstract: A large part of Australia's broad acre irrigation industry, which includes cotton, is farmed on heavy clay Vertosols. Recent changes in nematicide chemical availability, changes in rotations and the observation of the reniform nematode in central Queensland has highlighted that we need to improve our understanding of nematodes in these soils. We undertook preliminary investigations into distribution by depth under a cotton-cotton and cotton-maize rotation as well as vertical movement experiments in microcosms to better understand nematode distribution and movement in heavy clay soils. Analysis revealed that field populations decreased with soil sample depth, but there were also differences between rotations. In microcosm experiments, vertical movement of nematodes in these heavy clay soils was restricted, even in the presence of plant roots and moisture, both of which were hypothesised to improve nematode migration. The results imply that crop rotation currently remains a plausible option for nematode control, and that we still have a lot to learn about the ecology of nematode populations in Vertosols.

Keywords: *Gossypium*; *Zea mays*; vertisol; reniform

---

1. Introduction

In 2007, several experiments were undertaken within the Namoi valley cotton production area of New South Wales (NSW), Australia. These experiments were looking for interactions between genetically modified cotton and the soil biota [1], as well as the potential for an interaction between nematodes and the verticillium wilt [2,3], which is a production issue in the valley. At that time, there was no known nematode issue affecting Australian cotton production, although some potentially pathogenic nematodes were isolated [4,5], but these were in low numbers and possibly controlled by flood irrigation and the use of aldicarb [6].

Changes in funding and relocation of staff meant that continued monitoring was not possible; however, in 2014, a reversal in circumstance meant sampling, albeit to a limited extent, was recommenced. During the break in monitoring several changes occurred in the production system [7], with the removal of aldicarb and a shift to rotations that included maize being of note [8,9]. Additionally, *Rotylenhus reniformis* had been associated with yield losses around the Theodore area of central Queensland [10], which acted as a reminder of the importance of the Australian cotton industries 'come clean, go clean' policies [11]. The impact of reniform in Theodore also highlighted an industry requirement for more information on our nematode populations if we were to attempt to avoid the issues that were experienced in the USA. In the USA, reniform spread across almost half of the cotton fields of Alabama, Louisiana, and Mississippi in 50 years, reducing the yields by up to 20% [12,13].

We asked two questions to address some of the current unknowns, with regard to the Australian cotton production system. One was whether the inclusion of maize into the cotton rotation could affect the distribution of nematodes in the soil profile? The second was, do nematodes have the potential to move up a soil profile under favourable conditions? We undertook a combination of field core assessments and glasshouse based recolonization studies to address these questions. The results of these experiments are presented and discussed.

2. Materials and Methods

2.1. Soil Sites and Characteristics

Vertical distribution of nematodes, with regard to rotation, was recovered from soils taken from field C1 at the Australian Cotton Research Institute (ACRI), Narrabri, NSW. The soil is an alkaline dark grey clay Vertosol (approximately 66% clay) with a known decreases in soil carbon down the profiles [14]. The rotation on the site has previously been explained in detail [15], and cores were taken to a depth of 1 m in January of 2017 with a portable coring rig [16] from within the cotton-cotton and cotton-maize rotations when both rotations were planted to cotton. Cores were returned to the University of New England (UNE), where they were divided into 0–15, 15–30, 30–50, 50–70, and 70–100 cm depths and nematodes were extracted using a passive recovery technique [17] prior to enumeration. Other field parameters, such as cropping history and planting dates, were gathered from field records at the time of sampling.

The soils gravimetric water content (GWC) was assessed by comparing the weight of a field fresh sample with the resultant weight after drying to a constant mass at 105 °C. The dry weight bulk density was calculated from the mass of the soils that were recovered from the core while assuming no compaction during sampling.

Two soils were used in the vertical movement experiments. The first, designated 'Kirby', was collected from UNE's Kirby farm and it was a sandy loam (grey Chromosol [18]); 73% sand, 12% silt, and 14% clay with a pH_{H2O} (1 to 5 in water) of 5.4. The second soil, 'Cotton', was collected from a cotton property near Moree, NSW and it was a clay soil (black Vertosol [18]); 9% sand, 16% silt, and 74% clay with a pH_{H2O} of 8.2.

2.2. Soil Sterilization for Vertical Movement

Soil was autoclaved in 1 kg amounts at 20% GWC in open bags for one hour at 121 °C, at 1.5 bar and with the process repeated three times, with a 24 h break between the commencements of each autoclave cycle. Upon the completion of the sterilisation process, the autoclaved aliquots were combined into a sterile polypropylene bag and then left for two weeks in an open aseptic environment. After this time, three samples were taken from the soil and screened for nematode presence using passive extraction.

2.3. Microcosm Design

The microcosms were made from an unplasticised polyvinyl chloride (uPVC) pipe with an internal diameter of 50 mm. The pipe was cut into 40 cm lengths, which were then cut longitudinally to allow for the microcosm to be split lengthwise to facilitate soil recovery. The bottom of the microcosm was held together and sealed with a 50 mm uPVC end cap and the top of the tube with a 50 mm uPVC pipe to pipe joining collar. The cut edges of the pipe were sealed with tape to prevent water loss and splitting under expansion of the soil. Under experimental conditions, the microcosms were supported in plastic crates, which carried up to 16 microcosms.

2.4. Microcosm Packing

The microcosms were packed, so that sterile and non-sterile soil was represented in all combinations within the experiments as either a top (0–15 cm) or bottom (15–30 cm) treatment. This meant that there was; Kirby top: Kirby bottom, Kirby sterile top: Kirby bottom, Kirby top: Kirby sterile bottom, and Kirby sterile top: Kirby sterile bottom with the same combinations for the Cotton soil. The soils were packed to generate a dry bulk density of 1.4 g/cm^3, which was achieved by weighing the required mass of soil for each half of the microcosm and adding one-third of the mass at a time before tamping the tube five times on the bench to get the required compaction. An internal 15 cm mark was present in each tube to assist with packing to the desired bulk density. After either the bottom or the tops of the tubes were packed water was added to the presenting surface to raise the gravimetric water content of the soil to 20%.

2.5. Planting and Watering

Into the planted microcosms two seeds of wheat, variety Gregory, were planted to a depth of 1 cm and then the tops of all the microcosms were overlaid with 20 mL of 4 mm polypropylene beads to reduce evaporation. The initial starting weight of each established microcosm was taken and the GWC maintained by weight every Monday, Wednesday, and Friday of the experiment duration with the addition of variable amounts of rainwater to within 0.25 g of starting weight.

In a second experiment, a flood irrigation for half of the planted and unplanted microcosms was conducted two weeks after establishment by adding 50 mL of rain water to each of the identified microcosms. This was calculated as being sufficient water to raise the GWC to 35%, which had been established as being equivalent to −10 kPa.

2.6. Recovery and Nematode Counting

The microcosms were destructively sampled four weeks (28 days) after sowing wheat. The above ground plant height was recorded and the plant shoot material excised. Fresh weight was determined and the samples were dried for 48 h at 80 °C to determine the dry weight. Plastic beads were recovered from the top of the microcosms and then the tape and top and bottom caps were removed. The microcosms were opened in a large tray and the depth of visible root growth recorded. Soil was then recovered from 5 to 10 cm and 20 to 25 cm depths. A proportion of this soil was recovered to an aluminium tray to determine the GWC and approximately 10 g was weighed into a 50 mL centrifuge tube for nematode recovery [17].

2.7. Results and Analysis

Excel was used to tabulate results and interrogate data for correlation coefficients (r) generation. GenStat was used to undertake analysis of variance (ANOVA) of the measured variables, with Tukey's comparison test used to determine differences between multiple means with significance assumed to occur at the $p < 0.05$ level. Outcomes were graphically presented.

3. Results

3.1. Vertical Distribution

The total free living nematode populations were observed to decrease with depth under both the cotton-cotton and cotton-maize rotations with the overall population decline fitting the equation $y = -0.0928x^3 + 0.8549x^2 - 2.7682x + 4.6508$, with a correlation of r = 0.99. There was no significant difference between the rotations ($p = 0.07$), but there was a difference with depth ($p = 0.001$). An interaction between depth and rotation ($p = 0.02$) was observed with a larger nematode population in the cotton-cotton rotation between 30 to 70 cm than that recovered from under the cotton-maize rotation (Figure 1).

There was a good correlation between soil gravimetric water and nematode recovery from the cotton-maize rotation (r = 0.87), but not for cotton-cotton (r = 0.28). Both of the systems had good correlation between soil bulk density and the average number of nematodes (r = 0.80 and 0.84), with nematode abundance following a negative exponential curve as the bulk density increased.

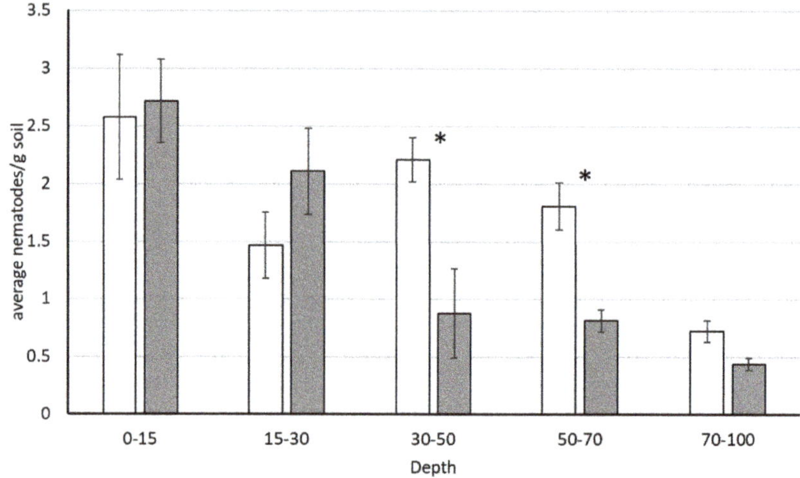

Figure 1. The mean free living nematodes per gram of soil recovered from a Vertosol supporting a cotton-cotton (white) and cotton-maize (grey) rotation to varying depths. Error bars represent the standard error of the means (n = 3) and the asterisks (*) indicate a significant difference between rotation and depth (p = 0.02).

3.2. Vertical Movement

The initial nematode populations were enumerated at 6.1 and 9.8 nematodes/g for the Kirby and Cotton soils, respectively. Examination of the soils, post sterilisation recovered no live nematodes in the Kirby soil, but the Cotton soil had 1.1 nematode/g (11% of the original nematode population) still alive after three rounds of autoclaving.

In both microcosm experiments, the wheat roots reached the bottom of the columns in the Kirby soil (30 cm), but only managed an average depth of 20.6 cm in the Cotton soil. Despite this, the wheat biomass was significantly higher (p < 0.001) in the Cotton soil than the Kirby soil, with means of 0.3 and 0.1 g, respectively.

In the first microcosm experiment, there was no significant difference in the nematode recovery between the Kirby and Cotton soils (p = 0.32), the top and bottom of the microcosms (p = 0.33), and whether wheat was planted or not (p = 0.11). Despite not being significant, nematode recovery, being expressed as a ratio of the control, implied movement up into sterile Kirby soil in both the presence and absence of wheat (Figure 2a). The average ratio of nematodes in sterile Cotton soil did not get above 1 in upper sterile Cotton soil, which implied a lack of upward movement (Figure 2a). In the bottom of the microcosms, there was a trend for increased nematode recovery in both sterile Kirby and Cotton soils, but only when wheat was planted (Figure 2a), despite the maintained 20% gravimetric water content.

In the second microcosm experiment, imposing flood irrigation on the Cotton soil significantly increased the number of recovered nematodes (p = 0.07), with 2.7 as compared to 1.63 nematodes/g for irrigated and GWC maintained soil, respectively. There was no significant difference in nematode recovery from either top or bottom of the microcosm (p = 0.39). Planting wheat had no significant effect on nematode recovery (p = 0.41), although the nematode recovery ratio increased above 1 for both irrigation treatments in the absence of planted wheat (Figure 2b).

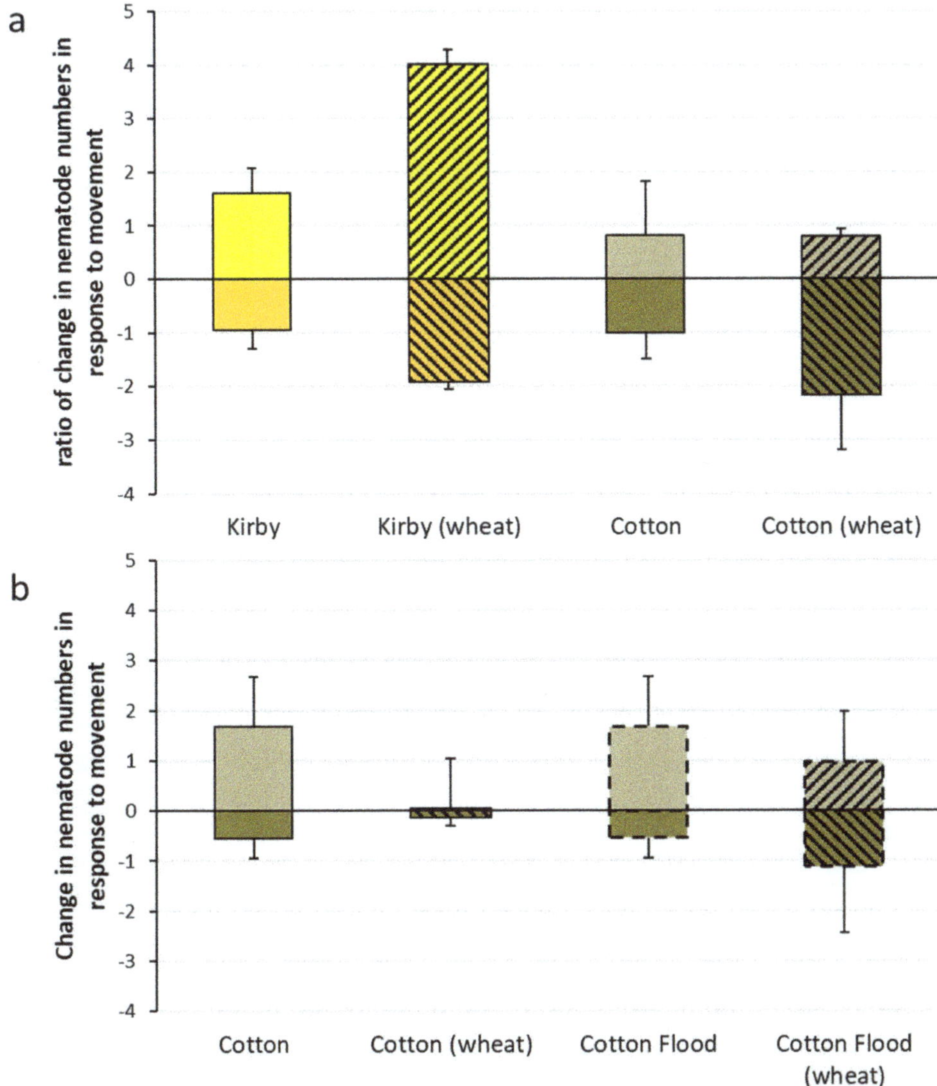

Figure 2. The ratio of nematodes recovered from sterilised soil situated either above or below non-sterile soil, compared to those recovered from a completely sterile treatment. A ratio of more than 1 (for upward movement, lighter shades) and −1 (for downward movement, darker shades) indicates an increase over the control. Kirby (Yellow) and Cotton (Brown) indicate where soil was sourced with (**a**) looking at the impact of sowing wheat (diagonal black shading) on nematode recovery, while (**b**) is the analysis of the impact of a maintained versus flood irrigation treatment (black dashed border) only in the Cotton soils.

4. Discussion

Farming systems are prone to change and the Australian cotton production system is no exception. However, the focus of these changes are often on either crop productivity or chemical and physical properties of the soil [19], with less attention being given to the soil biology [1], despite the fact that most, if not all, of our production diseases and pests are biological. We attempted to address some simple questions relating to nematodes in these systems in the face of the first observations of reniform nematode causing problems in Australian cotton [10] and the loss of potential chemical controls [8].

Our initial focus was on whether nematode populations declined with depth and whether rotations could influence their distribution. Our results indicated that there was a decline with depth as well as differences between rotations. Given that we sampled at a time when both of the rotations were growing cotton, we believe it would be safe to assume that the dissimilarity in the recovered nematodes/g between 30 to 70 cm (Figure 1) occurred due to rotational difference. With cotton being a tap rooted eudicot and maize a fibrous rooted monocot, a probable driver for changes in the nematode numbers between these depths is rooting patterns [20,21] in combination with these roots persisting post-harvest [22]. Root exudation and decomposition both have the potential to alter the soil microbiology [23], which, in turn, would directly influence both the nematode community composition and size [24]. Differences in the field management that are associated with the different rotational crops, such as fertilizer regimes, cultivation, and stubble management, could also be altering the soil microbial community and in turn the nematodes [25]. In keeping with this, cotton and maize roots are known to differentially alter the soils' abiotic properties [9], thus potentially altering the nematode population densities, which was supported with the observed correlations between nematode numbers, soil moisture, and bulk density. What a change is abundance does not address is whether it is also associated with a change in the population's trophic groups? Unfortunately, limitations on the volume of soil in our microcosms, our inability to remove all of the nematodes from the Cotton soil with autoclaving and the recoveries of only one to two nematodes/g from the recolonized soil, there was insufficient numbers to confirm this. However, with known pathogenic nematodes in these soils and a potential industry threat identified elsewhere, the difference in nematode abundance in soil from under the different rotations adds support for rotational crops remaining one of the few strategies available at present for nematode control in Australian cotton system [26,27].

Having observed a difference between the rotations, we postulated whether there was potential for nematodes to move vertically within these soils. Vertical nematode movement has been previously reported, notably for several plant parasitic nematodes that recolinise and recover from populations that reside deeper in the soil after crop protection control measures, such as nematicide application, have been implemented [28,29]. However, this work was undertaken on lighter soils than the Vertosol soil being investigated here [29]. Water is known to play a key role in both nematode movement and shaping community structure [30–32], and so we initially kept our soils at a moisture level that should have facilitated nematode movement [30]. However, in our limited and short term experiments, nematode movement either up or down in a heavy clay Vertosol appeared to be restrictive (Figure 2). In addition, we included the planting of wheat as a treatment factor, while assuming that the presence of growing roots might encourage nematode movement [33], but we observed no significant movement in response to plant roots (Figure 2). While surprising, it has been previously reported that the vertical distribution of roots does not always correlate to nematode movement or abundance [34]. While our microcosm experiments imply limited nematode movement and recolonisation potential in Vertosols, there are a number of caveats to consider prior to deriving any generalisations regarding nematode movements in these heavy clay soils. Firstly, our system was only run for four weeks, a relatively short period of time in a cropping cycle, we had limited replication and our Vertosol columns were not exposed to repeated flooding and drying cycles, as experienced under field conditions, but kept constantly moist. Finally, we did not work on the soils containing the reniform nematode due to quarantine concerns, but, given the potential for nematodes to behave differently, could not rule out the potential for *R. reniformis* to recolonise Vertosols from depth after flooding [10,28].

Accordingly, whilst these studies were preliminary, it is apparent that we still have much to learn about the diversity, potential threats, activity and importance of nematodes in Australian Vertosols, which themselves are challenging to work with. Within these heavy clay soils, the potential to use crop selection as a control strategy remains [26,27]. In the face of a reduction in available chemical controls [8], this strategy may continue to be one of the few mitigation options other than preventing nematode movement from infected fields [10] in the first place by maintaining 'come clean, go clean' practices [11].

Author Contributions: O.K. and B.G. undertook sampling, analysis and manuscript preparation. K.P. assisted with soils analysis and E.F. assisted with soil sampling and nematode enumeration. All authors have read and agreed to the published version of the manuscript.

Funding: This work was made possible with funding from the CRDC under UNE1403 and UNE2001. The University of New England (UNE) contributed via the GRASS programme, which supported Elijha, Katherine's PhD and in-kind and cash to UNE1403 and UNE2001.

Acknowledgments: Access to the trial sites and farms is gratefully acknowledged as is glasshouse support from Mick Faint at UNE.

Conflicts of Interest: The authors declare no conflict of interest.

References

1. Knox, O.G.G.; Gupta, V.V.S.R.; Lardner, R. Field evaluation of the effects of cotton variety and GM status on rhizosphere microbial diversity and function in Australian soils. *Soil Res.* **2014**, *52*, 203–215. [CrossRef]
2. Francl, L.J.; Wheeler, T.A. Interaction of Plant-Parasitic Nematodes with Wilt-Inducing Fungi. In *Nematode Interactions*; Khan, M.W., Ed.; Springer: Dortrecht, Germany, 1993; pp. 79–103.
3. Katsantonis, D.; Hillocks, R.J.; Gowen, S. Comparative Effect of Root-Knot Nematode on Severity of Verticillium and Fusarium Wilt in Cotton. *Phytoparasitica* **2003**, *31*, 154–162. [CrossRef]
4. Knox, O.G.G.; Anderson, C.M.T.; Allen, S.J.; Nehl, D.B. Helicotylenchus Dihystera in Australian Cotton Roots. *Australas. Plant Pathol.* **2006**, *35*, 287–288. [CrossRef]
5. Knox, O.G.G.; Anderson, C.M.T.; Nehl, D.B.; Gupta, V.V.S.R. Observation of Tylenchorhynchus ewingi in association with cotton soils in Australia. *Australas. Plant Dis. Notes* **2006**, *1*, 47–48. [CrossRef]
6. Cox, C. Aldicarb. *J. Pestic. Reform* **1992**, *12*, 4.
7. Braunack, M.V. Cotton farming systems in Australia: Factors contributing to changed yield and fibre quality. *Crop Pasture Sci.* **2013**, *64*, 834–844. [CrossRef]
8. Erickson, B. Pesticides: Bayer Cropscience, EPA Agree to Phase Out Use of Aldicarb. *Chem. Eng. News* **2010**, *88*, 1. [CrossRef]
9. Hulugalle, N.R.; McCorkell, B.; Heimoana, V.F.; Finlay, L.A. Soil properties under cotton-corn rotations in Australian cotton farms. *J. Cotton Sci.* **2016**, *20*, 294–298.
10. Bauer, B.; Smith, L.; Scheikowski, L.; Lehane, J.; Cobon, J.; O'Neill, W. Reniform Nematode Surveys in Central Queensland Cotton. In Proceedings of the 17th Australian Cotton Conference, Gold Coast, Australia, 4–6 August 2020; pp. 1–3.
11. Maas, S. Come Clean Go Clean. *Cottoninfo Fact Sheet* **2014**, 2. Available online: http://www.insidecotton.com/xmlui/handle/1/1015 (accessed on 4 February 2020).
12. Gazaway, W.S.; Mclean, K.S. A survey of plant-parasitic nematodes associated with cotton in Alabama. *J. Cotton Sci.* **2003**, *7*, 1–7.
13. Monfort, W.S.; Kirkpatrick, T.L.; Mauromoustakos, A. Spread of Rotylenchulus reniformis in an Arkansas Cotton Field Over a Four-Year Period. *J. Nematol.* **2008**, *40*, 161–166.
14. Polain, K.; Guppy, C.; Knox, O.; Lisle, L.; Wilson, B.; Osanai, Y.; Siebers, N. Determination of Agricultural Impact on Soil Microbial Activity Using δ18OP HCl and Respiration Experiments. *ACS Earth Space Chem.* **2018**, *2*, 683–691. [CrossRef]
15. Nachimuthu, G.; Hulugalle, N.R.; Watkins, M.D.; Finlay, L.A.; McCorkell, B. Irrigation induced surface carbon flow in a Vertisol under furrow irrigated cotton cropping systems. *Soil Tillage Res.* **2018**, *183*, 8–18. [CrossRef]
16. Polain, K.; Joice, G.; Jones, D.; Pereg, L.; Nachimuthu, G.; Knox, O.G.G. Coring lubricants can increase soil microbial activity in Vertisols. *J. Microbiol. Methods* **2019**, *165*, 105695. [CrossRef] [PubMed]
17. Knox, O.G.G.; Griffiths, B.S. Refinement of Passive Nematode Recovery from Cotton Growing High Clay Content Australian Vertisols. *Commun. Soil Sci. Plant Anal.* **2017**, *48*, 316–325. [CrossRef]
18. Isbell, R.F. *The Australian Soil Classification*; CSIRO Publishing: Collingwood, Australia, 1996.
19. Hulugalle, N.R.; Nehl, D.B.; Weaver, T.B. Soil properties, and cotton growth, yield and fibre quality in three cotton-based cropping systems. *Soil Tillage Res.* **2004**, *75*, 131–141. [CrossRef]

20. Hulugalle, N.R.; Broughton, K.J.; Tan, D.K.Y. Root growth of irrigated summer crops in cotton-based farming systems sown in Vertosols of northern New South Wales. *Crop Pasture Sci.* **2015**, *66*, 158–167. [CrossRef]

21. Luelf, N.; Tan, D.; Hulugalle, N.; Knox, O.; Weaver, T.; Field, D. Root turnover and microbial activity in cotton farming systems. In Proceedings of the 13th Australian Agronomy Conference, Perth, Australia, 10–14 September 2006.

22. Machinet, G.E.; Bertrand, I.; Chabbert, B.; Watteau, F.; Villemin, G.; Recous, S. Soil biodegradation of maize root residues: Interaction between chemical characteristics and the presence of colonizing micro-organisms. *Soil Biol. Biochem.* **2009**, *41*, 1253–1261. [CrossRef]

23. Broeckling, C.D.; Broz, A.K.; Bergelson, J.; Manter, D.K.; Vivanco, J.M. Root exudates regulate soil fungal community composition and diversity. *Appl. Environ. Microbiol.* **2008**, *74*, 738–744. [CrossRef]

24. Ferris, H.; Griffiths, B.S.; Porazinska, D.L.; Powers, T.O.; Wang, K.-H.; Tenuta, M. Reflections on Plant and Soil Nematode Ecology: Past, Present and Future. *J. Nematol.* **2012**, *44*, 115–126.

25. Griffiths, B.S.; Philippot, L. Insights into the resistance and resilience of the soil microbial community. *FEMS Microbiol. Rev.* **2013**, *37*, 112–129. [CrossRef]

26. Roberts, P.A. Current status of the availability, development, and use of host plant resistance to nematodes. *J. Nematol.* **1992**, *24*, 213. [PubMed]

27. Stirling, G.R. Biological control of plant-parasitic nematodes. In *Diseases of Nematodes*; Poinar, G.O., Ed.; CRC Press: Boca Raton, FL, USA, 2018; pp. 103–150.

28. Lee, H.K.; Lawrence, G.W.; DuBien, J.L.; Lawrence, K.S. Seasonal variation and cotton-corn rotation in the spatial distribution of *Rotylenchulus reniformis* in Mississippi cotton soils. *Nematropica* **2015**, *45*, 72–81.

29. Pinkerton, J.N.; Mojtahedi, H.; Santo, G.S.; O'Bannon, J.H. Vertical Migration of Meloidogyne chitwoodi and M. hapla under Controlled Temperature. *J. Nematol.* **1987**, *19*, 152–157. [PubMed]

30. Knox, O.G.G.; Killham, K.; Artz, R.R.E.; Mullins, C.; Wilson, M. Effect of Nematodes on Rhizosphere Colonization by Seed-Applied Bacteria. *Appl. Environ. Microbiol.* **2004**, *70*, 4666–4671. [CrossRef] [PubMed]

31. Liu, M.; Chen, X.; Qin, J.; Wang, D.; Griffiths, B.; Hu, F. A sequential extraction procedure reveals that water management affects soil nematode communities in paddy fields. *Appl. Soil Ecol.* **2008**, *40*, 250–259. [CrossRef]

32. Wallace, H.R. The Dynamics of Nematode Movement. *Annu. Rev. Phytopathol.* **1968**, *6*, 91–114. [CrossRef]

33. Standing, D.; Knox, O.G.G.; Mullins, C.E.; Killham, K.K.; Wilson, M.J. Influence of Nematodes on Resource Utilization by Bacteria; an in Vitro Study. *Microb. Ecol.* **2006**, *52*, 444–450. [CrossRef]

34. Forge, T.A.; DeYoung, R.; Vrain, T.C. Temporal changes in the vertical distribution of Pratylenchus penetrans under raspberry. *J. Nematol.* **1998**, *30*, 179.

Article

Response of the Arthropod Community to Soil Characteristics and Management in the Franciacorta Viticultural Area (Lombardy, Italy)

Isabella Ghiglieno [1], Anna Simonetto [1,*], Francesca Orlando [1], Pierluigi Donna [2], Marco Tonni [2], Leonardo Valenti [3] and Gianni Gilioli [1]

[1]   Department of Molecular and Translational Medicine, University of Brescia, 25123 Brescia, Italy;
      i.ghiglieno@unibs.it (I.G.); francesca.orlando@unibs.it (F.O.); gianni.gilioli@unibs.it (G.G.)
[2]   Sata Studio Agronomico, 25121 Brescia, Italy; pierluigi.donna@agronomisata.it (P.D.);
      marco.tonni@agronomisata.it (M.T.)
[3]   Department of Agricultural and Environmental Sciences—Production, Landscape, Agroenergy,
      University of Study of Milan, 20133 Milan, Italy; leonardo.valenti@unimi.it
*   Correspondence: anna.simonetto@unibs.it

Received: 12 April 2020; Accepted: 18 May 2020; Published: 21 May 2020

Abstract: Soil represents an important pool of biodiversity, hosting about a quarter of the living species on our planet. This soil richness has led to increasing interest in the structural and functional characteristics of its biodiversity. Studies of arthropod responses, in terms of abundance and taxon richness, have increased in relation to their ecological value as bioindicators of environmental change. This research was carried out over the 2014–2018 period with the aim to better understand arthropod taxa responses in vineyard soils in Franciacorta (Lombardy, Italy). To determine the biological composition in terms of arthropod taxa presence, one hundred soil samples were analysed. Environmental characteristics, such as chemical composition, soil moisture and temperature, and soil management were characterized for each soil sample. A total of 19 taxa were identified; the NMDS model analysis and the cluster analysis divided them into five groups according to their co-occurrence patterns. Each group was related to certain abiotic conditions; of these, soil moisture, temperature and organic matter were shown to be significant. A decision tree analysis showed that a longer period since conversion from conventional to organic farming lead to a higher arthropod biodiversity defined as a higher number of taxa.

Keywords: soil biodiversity; vineyard; co-occurrence patterns; soil moisture; soil temperature; soil organic matter; soil pH; vineyard management

---

1. Introduction

Soil has recently been described as the most complex and diverse ecosystem in the world [1], and it represents an important pool of biodiversity. It is indeed one of the richest habitats of terrestrial ecosystems in terms of species diversity [2,3]. The European Commission [4] estimates that about a quarter of living species on our planet are found in the soil, and the importance of this biodiversity has already been described in relation to the functional roles that the soil biota plays in regulating ecosystem processes [5]. However, despite the increasing number of studies on soil biodiversity, many structural and functional aspects of this biodiversity remain largely unexplored [5,6]. In this context, the investigation of the relationship between soil arthropod communities, in terms of abundance and taxa richness, and environmental conditions played an important role. The sensitivity of soil arthropods to environmental conditions [7,8], soil properties [9] and soil management practices [10] allows them to be considered as bioindicators of environmental change [11,12].

The relative importance of the factors influencing soil arthropod diversity and abundance in agroecosystems is still far from being understood. Indeed, the influence of abiotic and biotic variables and their interactions [13,14] varies according to the climate, type of soil and agricultural practices. The influence of meteorological variables (i.e., precipitation and air temperature), soil moisture and temperatures on soil arthropods has already been evaluated in different habitats [15–23]. Soil moisture and soil temperature have emerged as important factors that determine arthropod distribution [16,19,21,24,25], but the response of soil arthropods to soil water availability and temperature has been shown to vary between taxa [26–29]. In general, the positive effect of soil moisture on the abundance of soil arthropod communities has been emphasised [16,30] and an optimum temperature range of between 5 °C and 10 °C was identified for species active in winter, and between 10 °C and 18 °C for those active in summer [31] (p. 6). Soil chemical and physical characteristics have been identified as important drivers in soil arthropod distribution and abundance [9,11,13,20,32–38]. Soil texture [39], soil organic matter content [40,41], pH [11,37,38] and heavy metal concentration [13,20] have been shown to have a major influence on soil biota. Of these, soil pH and soil organic matter represent the most significant drivers in relation to the influence of soil pH variation on soil arthropods presence [11,37] and to arthropods role in soil organic matter degradation [34,35]. Soil arthropods contribute, in fact, to nutrient cycling as secondary decomposers, conditioning litter through comminution and passage through the gut, for further breakdown by the microflora [34] and stimulating microbial mineralisation of nutrients through grazing activity [35].

The influence of management on soil arthropods has been investigated in different agricultural contexts [10,42–46]. In particular, different studies have focused on the effect of organic viticulture on soil arthropod communities; different authors [15,47,48] show the general positive effect of organic management on soil arthropod abundance and distribution. However, the results have varied for each taxon investigated [49,50]. Furthermore, only few studies have been carried out evaluating the role of time of organic practice application on soil arthropod biodiversity [51]. Further research is therefore needed to assess the medium and long-term effects of organic agriculture on soil biodiversity [52].

In this paper, we report on the results of a 5-year investigation into the responses of the arthropod community to soil characteristics and vineyard management in the Franciacorta viticultural area (Lombardy, Italy). The diversity and co-occurrence patterns of different taxa were analysed in relation to abiotic factors, such as soil temperature, soil moisture and soil chemical properties. Moreover, the influence of vineyard management (conventional vs. organic) and the time of conversion from conventional to organic on arthropod biodiversity was investigated.

2. Materials and Methods

2.1. Study Sites

This study was carried out in a major Italian winemaking area. Franciacorta is the most famous Italian wine region for the production of sparkling wine using the champenoise method and is located in the Lombardy Region (Figure 1). The zone covers a total area of 2615 ha (as of 2018) and hosts 117 wineries (as of 2019). This research collected a total of 100 soil samples from 100 different vineyards over the period 2014–2018. Eighty-five per cent of samples were collected in spring (May or June) and 15% of samples in autumn (September, October or November). All the samples were characterised by presence of arthropods and the chemical characteristics of soil.

In Figure 1, the location of each sampling site is shown.

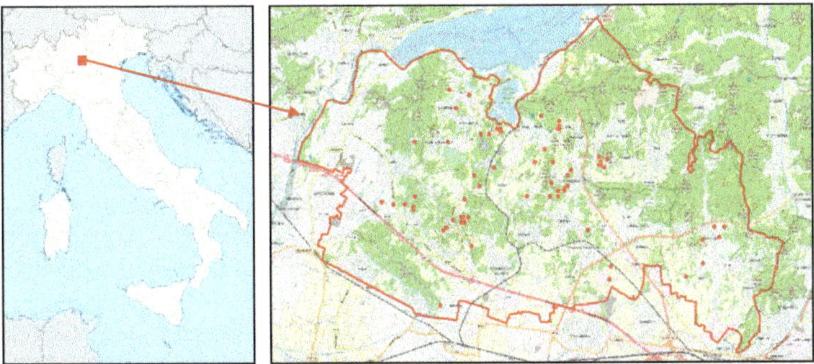

Figure 1. Map of the Franciacorta DOCG (Designation of Controlled and Guarantee Origin) winegrowing area. The locations of vineyards where samples were collected are indicated with red dots.

Vineyard management systems were classified in two main groups: conventionally managed vineyards without any specific environmental certification (conventional) and organic vineyards managed in compliance with the European Regulation on organic farming (reg EC n. 2018/848 and subsequent amendments and additions) (organic). For organic vineyards, we refer to the presence of this certification that implies compliance with the provisions of the law. In addition to this, we have verified a minimal set of conditions occurred in each farm monitored in organic farming. These actions refer to: no use of synthetic chemicals for plant protection and for fertilizing the vineyard; the integration of organic matter into the soil through the supply of organic matrices; the total absence of use of herbicides and the management of the sub-row through mechanical intervention; the preservation of the herbaceous covering on the ground; the minimum tillage adoption. Organic vineyards were then further divided into three subgroups, on the basis of how long ago they had been converted from conventional to organic farming: 3 years or less (organic ≤ 3), between 4 and 9 years (4 ≤ organic ≤ 9), and 10 years or more (organic ≥ 10).

2.2. Environmental Variables

Soil moisture (SM) and soil temperature (ST) data for the Franciacorta area from 2014 to 2018 were obtained from the National Centers for Environmental Predictions [53]. These data were then re-analysed using the Weather Research and Forecasting (WRF) simulations [54]. The WRF model (version 4.02) was applied to a high spatial resolution grid (each cell of the grid representing a 2×2 km area) to generate hourly data. In particular, the Noah scheme [55] has been used as land surface model (LSM) scheme (i.e., Noah, Noah-MP, and CLM4) to assess detailed multi-layer soil moisture and soil temperature. We focused on a depth of 0–15 cm below land surface. Each vineyard was associated with the nearest grid node to allow extraction of the specific soil temperature and moisture values.

To assess the influence of environmental variables on the presence of soil arthropods, SM and ST were evaluated for each vineyard in a 30-days reference period prior to the sampling date (Table A1). Two thermal thresholds (ST_{low} and ST_{up}) were considered to define two intervals of temperature. These intervals characterise organisms that prefer lower temperature features (taxa occurring more frequently in the interval $[ST_{low}, ST_{up}]$) or higher temperature features (taxa occurring more frequently when soil temperature is higher than ST_{up}). The lower threshold (ST_{low}) was set at 10 °C and the upper threshold (ST_{up}) was set at 20 °C [31] (p. 6). Soil temperatures lower than 10 °C were not included since they can be considered to be below the lower development threshold for most taxa. Considering ST_{low} and ST_{up}, two variables related to soil temperature were calculated for the reference period (720 h):

$$TL = \frac{1}{24} \sum_{i=1}^{720} (ST_i - ST_{low}), \ ST_{low} \leq ST_i < ST_{up} \tag{1}$$

$$TH = \frac{1}{24} \sum_{i=1}^{720} \left(ST_i - ST_{up} \right), \ ST_i \geq ST_{up} \tag{2}$$

where ST_i is hourly soil temperature. TL is the daily cumulative soil temperature degrees exceeding 10 °C when ST_i is between 10 °C and 20 °C; TH is the daily cumulative soil temperature degrees exceeding 20 °C when ST_i is greater than 20 °C.

A soil moisture threshold (SM) was defined to discriminate between organisms that prefer drier conditions, i.e., taxa occurring more frequently when soil moisture ranges in the interval [0, SM], or wetter conditions, i.e., taxa that more frequently occur when soil moisture is in the range [SM,1]. SM was set equal 0.35 (corresponding to 35%), which represents a reference value that can be associated, in different ecological contexts, to a status that satisfies the requirements in terms of humidity of soil arthropods [21,22]. Two variables associated with soil moisture were calculated for the reference period (720 h):

$$MD = \frac{1}{24} \sum_{i=1}^{720} |SM_i - SM|, 0 \leq SM_i \leq SM \tag{3}$$

$$MH = \frac{1}{24} \sum_{i=1}^{720} (SM_i - SM), \ SM_i > SM \tag{4}$$

where SM_i is the hourly soil moisture. MD is the daily sum of absolute deviations in soil moisture values from the threshold value when SM_i is lower than 0.35; MH is the daily cumulative soil moisture exceeding 0.35, when SM_i is higher than 0.35.

2.3. Chemical Characterisation of Soils

Chemical analysis of soils was performed according to the Italian regulation (DM 13 September 1999). Soil samples were taken at a depth of 0–15 cm and mixed homogeneously. Leaf litter was excluded, as it is not part of the soil itself. The collected soil samples were air-dried, homogenized and passed through a 2 mm sieve for chemical analysis.

Characterisation of the soil chemistry involved measuring soil texture (TXT), pH, active limestone (expressed in g $CaCO_3$/kg of soil) (AL), organic matter content (expressed in g/kg of soil) (SOM), available phosphorus (mg P_2O_5/kg of soil) (P), available potassium (mg K_2O/kg of soil) (K), available magnesium (mg MgO/kg of soil) (Mg) and copper content (mg/kg) (Cu). Soil texture was classified following the USDA soil texture triangle classification [56] (p. 125).

2.4. Soil Arthropods Identification

A cubic sample of soil (with a dimension of about 30 cm^3) was collected at the same depth described for chemical soil analysis, at each vineyard. Arthropods were extracted by placing the soil sample in a Berlese–Tüllgren funnel under a 60 W incandescence bulb, leading soil arthropods to migrate towards the damp part of the soil sample (away from the light). The soil arthropods fell through the cavity, into a preserving solution (2/3 alcohol and 1/3 glycerol). Determination of biological forms was carried out according to the QBS-ar (Soil Biological Quality-arthropod) method as proposed by [57], and the definition of the taxonomic entities and the biological stages is in agreement with the one reported in the same paper.

2.5. Data Analysis

2.5.1. Taxa Co-Occurrence Patterns

To measure soil arthropods biodiversity a taxa co-occurrence approach was used. For each vineyard, a taxa presence profile was defined, i.e., a vector indicating the presence or absence of the taxa in each vineyard. The presence profile did not consider population abundance. Based on the presence profiles, vineyards and taxa were described in a J dimensional space (J is the number of taxa considered), allowing taxa to be ordered by their vineyard presence profiles. Two taxa are close to each other if they share a similar pattern of co-occurrence in the vineyards, they are far from each other if one is present in the vineyards where the other is absent and vice versa.

To allow easy visualisation and interpretation of dissimilarity in soil biodiversity and taxa co-occurrence, it is useful to represent these profiles in a two-dimensional space, called an ordination plane. Non-metric multidimensional scaling (NMDS) can be used to summarise information and reduce the dimensionality of profiles [58]. By applying NMDS, vineyards and taxa can be ordered by the dissimilarity of the presence profiles. Bray–Curtis dissimilarity [59], used extensively in the ecological field, was adopted. NMDS analysis was performed using the metaMDS function of the vegan package in R [60]. Loss of information due to a reduction in dimensionality is assessed by the stress value, which refers to the disagreement between 2-D representation and original positions of taxa in multidimensional space.

To test which environmental drivers (Cu, pH, AL, SOM, P, K, Mg, TL, TH, MD and MH) are significantly correlated to the first two axes of the NMDS ordination plane, we applied the envfit function of the vegan R package [60]. Each variable was correlated independently and plotted on the plane as a vector. The direction of the vector represents the gradient direction of the environmental driver, while the length of the vector is proportional to the correlation of the ordination system and the environmental driver.

Taxa were grouped into clusters as homogeneous as possible in terms of co-occurrence patterns, based on taxa ordination results [59]. To perform hierarchical cluster analysis, the hclust function of R software [61] was applied.

2.5.2. Vineyard Management Impact

To assess the impact of vineyard management on the biodiversity of soil biota, decision tree analysis was performed. The number of taxa present in each soil sample was considered as a measurement of edaphic biodiversity, and three categories of soil biodiversity were defined: 'low' when the number of taxa was lower or equal to 4, 'medium' when the number of taxa in the soil sample was between 5 and 8, and 'high' when the number of taxa was greater than 8. A classification decision tree allowed to split the soil samples into homogeneous groups according to edaphic biodiversity based on the different vineyard management classes. Recursive partitioning and regression tree (RPART) analysis were performed by applying the rpart package of R software [62]. The fitting of the model was investigated using the accuracy index that corresponds to the percentage of cases correctly classified.

3. Results

3.1. Descriptive Analysis

3.1.1. Environmental and Vineyard Management Variables

The descriptive statistics for environmental variables included in the full model are shown in Table 1.

Seven types of soil texture were considered: clay, clay loam, silty clay loam, sandy clay loam, loam, silt loam and sandy loam.

Vineyard management was categorised into four classes: conventional management (7% of the sample), vineyards converted to organic farming in the last three years (45% of the sample), vineyards converted between 4 and 9 years ago (31% of the sample), and vineyards converted at least 10 years ago (17% of the sample).

Table 1. Descriptive statistics of continuous variables (soil characteristics and environmental drivers) included in the analysis.

	Unit of Measure	Mean ± Standard Deviation	Median (Q25–Q75)	Min	Max
Cu	(mg/kg)	58.68 ± 32.81	55.40 (36.9–72.2)	4.20	170.00
pH		7.10 ± 0.87	7.30 (6.35–7.9)	5.30	8.20
AL	(g $CaCO_3$/kg)	10.16 ± 22.62	0.00 (0.00–13.50)	0.00	130.00
SOM [1]	(g/kg)	21.94 ± 9.08	23.00 (15.00–25.00)	5.00	42.00
P	(mg P_2O_5/kg)	54.47 ± 40.20	51.00 (26.00–64.00)	9.00	222.00
K	(mg K_2O/kg)	148.52 ± 67.53	145.00 (94.00–178.00)	60.00	354.00
Mg	(mg MgO/kg)	165.75 ± 75.37	138.00 (117.00–210.00)	66.00	383.00
TL [2]	°C	68.76 ± 44.90	69.75 (33.67–104.39)	0.00	161.46
TH [3]	°C	106.05 ± 59.06	121.76 (55.73–153.38)	14.20	241.77
MD [4]	Pure number	10.39 ± 26.03	5.16 (1.51–7.88)	0.00	135.53
MH [5]	Pure number	0.11 ± 0.20	0.00 (0.00–0.156)	0.00	0.63

[1] SOM: soil organic matter. [2] TL: daily cumulative soil temperature exceeding 10 °C when soil temperature is between 10 and 20 °C. [3] TH: daily cumulative soil temperature exceeding 20 °C when soil temperature is higher than 20 °C. [4] MD: daily sum of absolute deviations in soil moisture when soil moisture is lower than 0.35. [5] MH: daily cumulative soil moisture exceeding 0.35 when soil moisture is higher than 0.35.

3.1.2. Taxa Identification

A total of 19 taxa were identified in the soil samples. In case of Diptera and Coleoptera, the biological stage of larvae were also detected (Table 2).

Collembola, Acari and Hymenoptera recorded the highest frequency of presence in the soil samples analysed. Collembola and Acari were reported in 89 of the 100 vineyards, Hymenoptera in 80 vineyards. The lowest frequency of occurrence was recorded for Psocoptera, Thysanoptera and Isopoda (8/100, 7/100, 6/100 respectively).

Table 2. Distribution of taxa according to stages considered in the analysis and presence (i.e., number of soil samples in which the taxon has been identified).

Taxa	Larvae	N° of Presences	Other Stages [1]	N° of Presences
Acari			x	89
Myriapoda—Diplopoda			x	12
Myriapoda—Chilopoda			x	17
Myriapoda—Symphyla			x	56
Myriapoda—Pauropoda			x	32
Hymenoptera			x	80
Thysanoptera			x	7
Pseudoscorpionida			x	11
Psocoptera			x	8
Coleoptera			x	31
Coleoptera larvae	x	39		
Collembola			x	89
Diptera			x	31
Diptera larvae	x	32		
Protura			x	25
Diplura			x	27
Hemiptera			x	12
Isopoda			x	6
Other_holometabolous [2]			x	20

[1] Other stages include all forms that produce active participation in soil cycles (e.g., pupae are excluded). In the case of the 'Other_holometabolous' taxon, the pupal stage is also included. [2] Other_holometabolous taxa include Mecoptera, Neuroptera and Raphidioptera orders in agreement with QBS-ar (Soil Biological Quality-arthropod) method [57].

3.2. Co-Occurrence Pattern Identification

Taxa dispersion in the non-metric multidimensional scaling plane is shown in Figure 2. Taxa were ordered according to their co-occurrence profiles. Neighbouring taxa in the plane were characterised by the presence in the same vineyards (e.g., Collembola and Coleoptera larvae, Psocoptera and Pseudoscorpionida); the more distant are two taxa, greater is the difference in terms of their presence in the vineyards (e.g., Diptera and Psocoptera, Acari and Pauropoda). The stress value estimated for the model was equal to 0.2, indicating the model has good ability to predict data in the reduced space.

The results of analysis of the correlation between environmental drivers and the NMDS plane are shown in Figure 2.

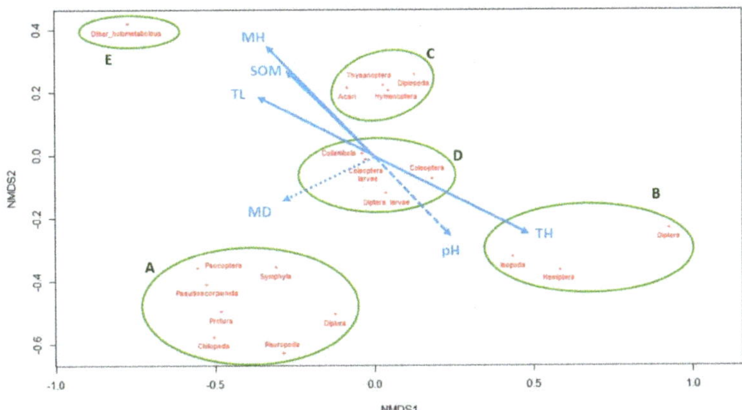

Figure 2. Results of non-metric multidimensional (NMDS) analysis: dispersion of taxa (red points) according to their co-occurrence profiles (NMDS1 and NMDS2 are the two axes of the ordination plane). Blue arrows refer to the correlation of environmental drivers and soil characteristics with NMDS ordination pattern (solid line ——— p-value < 0.5, dashed line – – · p-value < 0.1, dotted line ········ p-value < 0.15). The five clusters of taxa according their presence pattern are highlighted with the green circles.

p-Values of the correlation coefficients were used to discriminate the intensity of the relationship between environmental drivers and the taxa ordering system (Table 3): strong correlation for SOM, TL, TH and MH (p-value < 0.05); medium intensity correlation for pH (p-value < 0.1); low intensity correlation for MD (p-value < 0.15). The other environmental drivers were not significantly correlated with the first two axes of the NMDS system.

The results obtained from NMDS and cluster analysis (Figure 3) allowed the taxa to be divided into five groups according to their co-occurrence pattern. The five clusters shown in the cluster dendrogram correspond to the clusters identified by the green circles in the NMDS plane (Figure 2).

Group A included the largest number of taxa and specifically the Pseudoscorpionida, Psocoptera, Protura, Diplura Chilopoda, Symphyla and Pauropoda. Group B was made up of Diptera, Hemiptera and Isopoda taxa, while the larval form of Diptera was located in group D, together with Coleoptera, both as larvae and other biologic forms, and Collembola. The Acari, Hymenoptera, Thysanoptera and Diplopoda taxa made up group C. Group E is only represented by the taxa defined as 'Other_holometabolous'.

Table 3. Correlation analysis of environmental drivers and soil characteristics with NMDS ordination pattern.

Variable	Squared Correlation Coefficient	*p*-Value [6] of Correlation Coefficient	
Cu	0.05	0.17	
pH	0.06	0.09	**
AL	0.02	0.48	
SOM [1]	0.08	0.05	***
P	0.02	0.43	
K	0.01	0.66	
Mg	0.01	0.64	
TL [2]	0.08	0.04	***
TH [3]	0.15	0.01	***
MD [4]	0.05	0.15	*
MH [5]	0.12	0.01	***
TXT	0.04	0.83	

[1] SOM: soil organic matter. [2] TL: daily cumulative soil temperature exceeding 10 °C when soil temperature is between 10 and 20 °C. [3] TH: daily cumulative soil temperature exceeding 20 °C when soil temperature is higher than 20 °C. [4] MD: daily sum of absolute deviations in soil moisture when soil moisture is lower than 0.35. [5] MH: daily cumulative soil moisture exceeding 0.35, when soil moisture is higher than 0.35. [6] * *p*-value < 0.15, ** *p*-value < 0.1, *** *p*-value < 0.05.

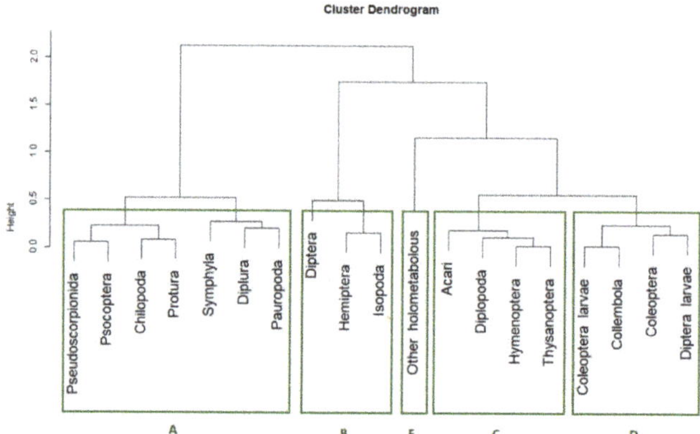

Figure 3. Dendrogram of hierarchical cluster analysis of taxa based on NMDS results. The five clusters are highlighted in green.

3.3. Vineyard Management

The results of the classification tree showed that variable vineyard management could be useful for discriminating different categories of soil biodiversity. In particular, the analysis of the tree shown in Figure 4 showed that conventionally managed vineyards were associated with a low level of biodiversity, vineyards that had adopted organic management for a maximum of three years were associated with a medium level of biodiversity, and vineyards that had adopted organic management for at least four years were associated with a high level of biodiversity.

The accuracy index showed a good fit of the model as 57% of cases was correctly classified.

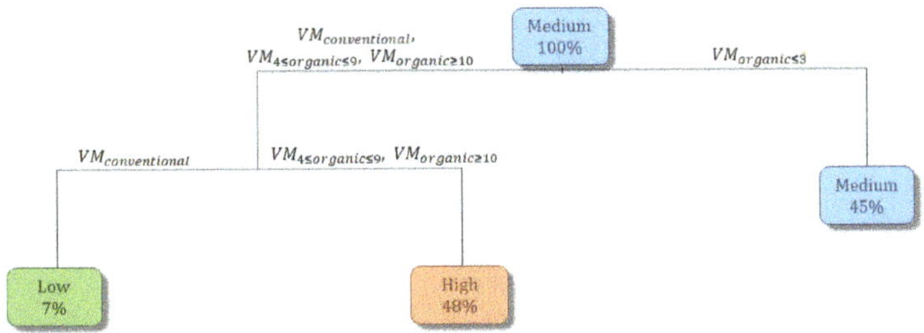

Figure 4. Classification decision tree of soil samples to predict soil biodiversity according to vineyard management. The predicted level of soil biodiversity (low, medium, high) is reported in the squared box, together with the percentage of soil samples included in that node. The paths from the initial box (with 100% of cases) to the final boxes represent the classification rules.

4. Discussion and Conclusions

The results obtained in this study allowed to identify the co-occurrence pattern for 19 taxa of soil arthropods on the basis of a 5-year investigation carried out in the Franciacorta viticultural area (Lombardy, Italy). The NMDS showed significant relationships between investigated soil arthropod taxa and soil moisture (MD and MH), soil temperature (TL, TH), soil organic matter (SOM) and pH. The decision tree showed an increased taxa diversity in relation to organic vineyard management and to the increase of time period of conversion from conventional to organic management.

In line with the expectations, Collembola and Acari were the most frequent of the 19 taxa identified, confirming that they are the most present groups of arthropods in soil [36,47,51]. The high level of presence of Hymenoptera recorded in our analysis is in agreement with other studies carried out in different agricultural contexts and reporting a significant presence of this taxon, mostly represented by Formicidae, in vineyard soils [63].

Based on taxa co-occurrence patterns, identified through NMDS analysis, five groups were found. Moreover, NMDS analysis made it possible to explore the relationship between soil abiotic variables and the aggregation of arthropod taxa in groups. In particular, the results obtained from our study pointed out that presence patterns characterising group A showed only one significant correlation (p-value < 0.15) with low soil moisture (MD). This result is compatible with the hypothesis that the taxa included in group A were relatively less dependent on high humidity values. The taxa in group B and the Coleoptera and Diptera larvae taxa (group D) were associated with higher pH (p-value < 0.1) and higher soil temperatures (TH) (p-value < 0.05), in line with the possible thermophilic habit of some representatives of these taxa [22,31,64]. The relationship with a higher pH level is more evident for the Isopoda and this is in agreement with van Straalen [11], who underlined weakly alkaliphilous or sub-neutral behaviour for some species of Isopoda. The detected ubiquitous presence of Collembola (group D) could be partially explained by the significant variability of responses to soil temperature, moisture and chemical properties of the different species of this taxon. In particular, the effect of soil moisture on Collembola has been documented by different authors [25,65], while species-specific responses have been reported [21]. Furthermore, Heiniger et al. [66] highlight that the role of microclimate for Collembola could be less important for their distribution than the role of trophic resources and competition. The presence of taxa in groups C and E is mostly determined by soil organic matter (SOM), soil moisture value higher than threshold level of 0.35 (MH) and lower temperature (TL) (p-values < 0.05). The relationship with SOM can be related to the involvement of these taxa in the soil food webs that starts from decomposition of dead organic matter generated by the activity of bacteria and fungi [32]. Diplopoda (Millipedes) are involved in SOM degradation, as their feeding activity is focused on dead organic matter [35,67]. A significant influence of soil nitrogen on species richness and biodiversity has been observed for this taxon [68], while Hymenoptera are involved in the decomposition of organic substances [35]. In relation to the positive response of group C to soil moisture increase, some authors have underlined that soil water availability is an important factor controlling presence of mites (Acari) [69]. Other authors have showed that Oribatid mites (Acari: Oribatida) are positively influenced by soil temperature [27] and that their distribution is dependent on soil moisture [35]. The relationship observed between group E and soil moisture can be related for Mecoptera (included in Other holometabolous taxon) with data reported for pre-imaginal stages of this order which develop in the soil and showed preference for high soil moisture [70].

The co-occurrence pattern of the taxa identified in our study is in line with similar pattern reported in the literature. Taxa co-occurrence in group B agrees with the results in [71] that confirmed Diptera and Isopoda co-existence in some specific habitats. Acari and Hymenoptera (group C) have also been grouped together by other authors [41]. The composition of groups A and C suggests that the co-occurrence pattern can also be influenced by biotic relationships among taxa. According to Eisenbeis and Wichard [31] (p. 192), the trophic niche of Diplura includes Symphyla, while Weygoldt [72] noted that Pseudoscorpionida feed on different orders of small soil arthropods, including Psocoptera. All these associations support the taxa co-occurrence in group A. Similarly, Coleoptera contain taxa (e.g., Carabid beetle) that have been described as predators of Collembola [73]. This association is in line with the co-occurrence of these two taxa in group D.

The results obtained analysing the role of vineyard management on soil arthropods diversity allows to identify an increase of taxa diversity in relation to organic vineyard management. This is consistent with previous studies, which reported a general increase of arthropod biodiversity [48] and arthropod abundance [47] associated to organic vineyard management. This effect was evident even before a 3-year period after conversion. The effect on arthropod biodiversity markedly increases with the length of the period since organic farming adoption.

The results obtained in this study provide additional knowledge supporting the interpretation of diversity and co-occurrence patterns in soil Arthropoda in vineyard. The importance of abiotic variables together with the interpretation of the possible role of biotic relationship among taxa have been explored in the specific geographic context of the Franciacorta viticultural area. Furthermore, our study confirmed the effect of organic vineyard management in increasing arthropod taxa diversity and, most importantly, it showed the critical role of the time of conversion from conventional to organic farming in increasing arthropod biodiversity. Further experiments are needed to extend these results to other viticultural contexts.

Author Contributions: Conceptualization, I.G., P.D., M.T., L.V. and G.G.; Data curation, I.G., A.S., F.O., M.T. and G.G.; Formal analysis, A.S.; Investigation, I.G., P.D., M.T. and L.V.; Methodology, I.G., A.S. and P.D.; Supervision, G.G.; Visualization, I.G. and A.S.; Writing—original draft, I.G. and A.S.; Writing—review & editing, G.G. All authors have read and agreed to the published version of the manuscript.

Funding: This research received no external funding.

Acknowledgments: The authors would like to thank all the wine companies assisting with data collection. We thank the 'Consorzio per la tutela del Franciacorta', which played a key role in the development of new operational protocols for sustainable viticulture by promoting and inspiring collaboration between universities and productive excellence. We thank the Chemical Laboratory and Organic Agriculture Unit of the Technology Transfer Center (CTT) of the Fondazione Edmund Mach (San Michele all'Adige, Trento, Italy), which carried out chemical analysis of soils and arthropods groups identification. We also thank TerrAria s.r.l. for providing high-resolution weather and soil data.

Conflicts of Interest: The authors declare no conflict of interest.

Appendix A

Table A1. Average values of soil temperature (ST °C) and soil moisture (SM measured in the interval 0–1) of the 30-days reference period prior to the sampling date for each site.

ID Field	Date	ST	SM	ID Field	Date	ST	SM
1	14 May	20.20	0.087	51	15 June	24.90	0.299
2	14 May	20.21	0.090	52	15 June	24.61	0.221
3	14 May	19.65	0.085	53	15 June	24.85	0.227
4	14 May	19.72	0.091	54	15 June	24.54	0.244
5	14 May	20.27	0.106	55	15 June	26.14	0.225
6	14 May	20.31	0.082	56	15 June	25.21	0.226
7	14 May	20.28	0.082	57	15 June	25.21	0.225
8	14 May	20.19	0.088	58	15 June	25.97	0.223
9	14 May	20.18	0.086	59	15 June	25.97	0.222
10	14 May	20.28	0.087	60	16 June	21.71	0.340
11	14 May	20.29	0.084	61	16 June	21.71	0.341
12	14 May	20.70	0.100	62	16 June	21.71	0.340
13	14 May	20.71	0.103	63	16 June	21.71	0.339
14	14 May	20.52	0.101	64	16 June	25.75	0.370
15	14 May	20.52	0.102	65	16 June	25.76	0.370
16	14 May	19.26	0.128	66	16 June	20.29	0.344
17	14 May	19.26	0.127	67	16 June	26.05	0.369
18	14 May	19.22	0.145	68	16 June	25.75	0.370
19	14 May	19.53	0.121	69	16 June	26.05	0.369
20	14 May	19.44	0.120	70	16 June	26.05	0.369
21	14 May	19.58	0.122	71	16 June	25.75	0.370
22	14 May	20.52	0.100	72	16 June	26.02	0.369
23	14 May	20.52	0.100	73	16 June	20.29	0.344
24	14 May	20.52	0.099	74	16 June	26.05	0.369
25	14 May	19.63	0.087	75	16 June	26.56	0.370
26	14 May	20.17	0.087	76	16 June	20.29	0.344
27	14 May	20.17	0.088	77	16 June	26.04	0.369
28	14 September	23.47	0.190	78	16 June	21.54	0.337
29	14 September	23.58	0.325	79	16 June	25.14	0.371
30	14 September	24.56	0.188	80	16 June	21.38	0.326
31	14 September	24.56	0.185	81	16 June	18.98	0.352
32	15 May	18.70	0.353	82	16 June	18.98	0.350
33	15 May	19.97	0.336	83	16 June	18.98	0.352
34	15 June	22.98	0.361	84	16 June	18.98	0.354
35	15 June	22.98	0.345	85	16 June	18.98	0.352
36	15 June	22.92	0.361	86	16 June	18.98	0.351
37	15 June	24.74	0.230	87	17 September	27.97	0.349
38	15 June	24.74	0.228	88	17 September	27.84	0.349
39	15 June	25.02	0.300	89	17 September	22.95	0.353
40	15 June	23.89	0.275	90	17 September	21.67	0.164
41	15 June	24.74	0.176	91	17 September	21.67	0.162
42	15 June	24.70	0.178	92	17 September	21.67	0.159
43	15 June	24.70	0.177	93	17 September	25.84	0.339
44	15 June	24.75	0.165	94	17 September	25.84	0.340
45	15 June	25.08	0.299	95	17 September	21.67	0.162
46	15 June	24.90	0.302	96	17 September	25.84	0.340
47	15 June	24.14	0.336	97	17 September	25.84	0.340
48	15 June	24.13	0.338	98	18 June	24.37	0.323
49	15 June	24.14	0.344	99	18 June	24.37	0.322
50	15 June	24.11	0.334	100	18 June	24.37	0.323

References

1. Kopittke, P.M.; Menzies, N.W.; Wang, P.; McKenna, B.A.; Lombi, E. Soil and the intensification of agriculture for global food security. *Environ. Int.* **2019**, *132*, 105078. [CrossRef] [PubMed]
2. Wolters, V. Biodiversity of soil fauna and its function. *Eur. J. Soil Biol.* **2001**, *37*, 221–227. [CrossRef]
3. Decaëns, T.; Jiménez, J.J.; Gioia, C.; Measey, G.J.; Lavelle, P. The values of soil animals for conservation biology. *Eur. J. Soil Biol.* **2006**, *42*, S23–S38. [CrossRef]
4. European Commission. *The Factory of Life. Why Soil Biodiversity Is so Important*; Office for Official Publications of the European Union: Luxembourg, 2010; p. 22.
5. Geisen, S.; Briones, M.J.I.; Gan, H.; Behan-Pelletier, V.M.; Friman, V.-P.; de Groot, G.A.; Hannula, S.E.; Lindo, Z.; Philippot, L.; Tiunov, A.V.; et al. A methodological framework to embrace soil biodiversity. *Soil Biol. Biochem.* **2019**, *136*, 107536. [CrossRef]
6. Cameron, E.K.; Martins, I.S.; Lavelle, P.; Mathieu, J.; Tedersoo, L.; Gottschall, F.; Guerra, C.A.; Hines, J.; Patoine, G.; Siebert, J.; et al. Global gaps in soil biodiversity data. *Nat. Ecol. Evol.* **2018**, *2*, 1042–1043. [CrossRef]
7. Juan-Ovejero, R.; Benito, E.; Barreal, M.E.; Rodeiro, J.; Briones, M.J.I. Tolerance to fluctuating water regimes drives changes in mesofauna community structure and vertical stratification in peatlands. *Pedobiologia* **2019**, *76*, 150571. [CrossRef]
8. Wallwork, J.A. Distribution Patterns and Population Dynamics of the Micro-Arthropods of a Desert Soil in Southern California. *J. Anim. Ecol.* **1972**, *41*, 291–310. [CrossRef]
9. Ruf, A.; Beck, L.; Dreher, P.; Hund-Rinke, K.; Römbke, J.; Spelda, J. A biological classification concept for the assessment of soil quality: "biological soil classification scheme" (BBSK). *Agric. Ecosyst. Environ.* **2003**, *98*, 263–271. [CrossRef]
10. Marasas, M.E.; Sarandón, S.J.; Cicchino, A.C. Changes in soil arthropod functional group in a wheat crop under conventional and no tillage systems in Argentina. *Appl. Soil Ecol.* **2001**, *18*, 61–68. [CrossRef]
11. van Straalen, N.M. Evaluation of bioindicator systems derived from soil arthropod communities. *Appl. Soil Ecol.* **1998**, *9*, 429–437. [CrossRef]
12. Büchs, W.; Harenberg, A.; Zimmermann, J.; Weiß, B. Biodiversity, the ultimate agri-environmental indicator?: Potential and limits for the application of faunistic elements as gradual indicators in agroecosystems. *Agric. Ecosyst. Environ.* **2003**, *98*, 99–123. [CrossRef]
13. Migliorini, M.; Pigino, G.; Bianchi, N.; Bernini, F.; Leonzio, C. The effects of heavy metal contamination on the soil arthropod community of a shooting range. *Environ. Pollut.* **2004**, *129*, 331–340. [CrossRef] [PubMed]
14. Xu, G.-L.; Kuster, T.M.; Günthardt-Goerg, M.S.; Dobbertin, M.; Li, M.-H. Seasonal exposure to drought and air warming affects soil Collembola and mites. *PLoS ONE* **2012**, *7*, 1–9. [CrossRef] [PubMed]
15. Ghiglieno, I.; Simonetto, A.; Donna, P.; Tonni, M.; Valenti, L.; Bedussi, F.; Gilioli, G. Soil Biological Quality Assessment to Improve Decision Support in the Wine Sector. *Agronomy* **2019**, *9*, 593. [CrossRef]
16. Frampton, G.; Van den Brink, P.; Gould, P. Effect of spring drought and irrigation on farmland arthropods in southern Britain. *J. Appl. Ecol.* **2001**, *37*, 865–883. [CrossRef]
17. Uvarov, A.V. Effects of diurnal temperature fluctuations on population responses of forest floor mites. *Pedobiologia* **2003**, *47*, 331–339. [CrossRef]
18. Jucevica, E.; Melecis, V. Global warming affect Collembola community: A long-term study. *Pedobiologia* **2006**, *50*, 177–184. [CrossRef]
19. Clapperton, M.J.; Kanashiro, D.A.; Behan-Pelletier, V.M. Changes in abundance and diversity of microarthropods associated with Fescue Prairie grazing regimes. *Pedobiologia* **2002**, *46*, 496–511. [CrossRef]
20. Holmstrup, M.; Maraldo, K.; Krogh, P.H. Combined effect of copper and prolonged summer drought on soil Microarthropods in the field. *Environ. Pollut.* **2007**, *146*, 525–533. [CrossRef]
21. Tsiafouli, M.A.; Kallimanis, A.S.; Katana, E.; Stamou, G.P.; Sgardelis, S.P. Responses of soil microarthropods to experimental short-term manipulations of soil moisture. *Appl. Soil Ecol.* **2005**, *29*, 17–26. [CrossRef]
22. Reddy, M.V.; Venkataiah, B. Seasonal abundance of soil-surface arthropods in relation to some meteorological and edaphic variables of the grassland and tree-planted areas in a tropical semi-arid savanna. *Int. J. Biometeorol.* **1990**, *34*, 49–59. [CrossRef]
23. Grear, J.S.; Schmitz, O.J. Effects of Grouping Behavior and Predators on the Spatial Distribution of a Forest Floor Arthropod. *Ecology* **2005**, *86*, 960–971. [CrossRef]

24. Choi, W.I.; Ryoo, M.I.; Kim, J.-G. Biology of Paronychiurus kimi(Collembola: Onychiuridae) under the influence of temperature, humidity and nutrition. *Pedobiologia* **2002**, *46*, 548–557. [CrossRef]

25. Choi, W.I.; Moorhead, D.L.; Neher, D.A.; Ryoo, M.I. A modeling study of soil temperature and moisture effects on population dynamics of Paronychiurus kimi (Collembola: Onychiuridae). *Biol. Fertil. Soils* **2006**, *43*, 69–75. [CrossRef]

26. O'Lear, H.A.; Blair, J.M. Responses of soil microarthropods to changes in soil water availability in tallgrass prairie. *Biol. Fertil. Soils* **1999**, *29*, 207–217. [CrossRef]

27. Liu, J.-L.; Li, F.-R.; Liu, L.-L.; Yang, K. Responses of different Collembola and mite taxa to experimental rain pulses in an arid ecosystem. *Catena* **2017**, *155*, 53–61. [CrossRef]

28. Ikemoto, T. Intrinsic Optimum Temperature for Development of Insects and Mites. *Environ. Entomol.* **2005**, *34*, 1377–1387. [CrossRef]

29. Ikemoto, T. Possible existence of a common temperature and a common duration of development among members of a taxonomic group of arthropods that underwent speciational adaptation to temperature. *Appl. Entomol. Zool.* **2003**, *38*, 487–492. [CrossRef]

30. Rapoport, E.H.; Tschapek, M. Soil water and soil fauna. *Rev. Ecol. Biol. Sol.* **1967**, *4*, 1–58.

31. Eisenbeis, G.; Wichard, W. *Atlas on the Biology of Soil Arthropods*; Springer: Berlin/Heidelberg, Germany, 1987.

32. Petersen, H.; Luxton, M. A Comparative Analysis of Soil Fauna Populations and Their Role in Decomposition Processes. *Oikos* **1982**, *39*, 288–388. [CrossRef]

33. Wu, P.; Liu, X.; Liu, S.; Wang, J.; Wang, Y. Composition and spatio-temporal variation of soil microarthropods in the biodiversity hotspot of northern Hengduan Mountains, China. *Eur. J. Soil Biol.* **2014**, *62*, 30–38. [CrossRef]

34. Culliney, T. Role of arthropods in maintaining soil fertility. *Agriculture* **2013**, *3*, 629–659. [CrossRef]

35. Bagyaraj, D.J.; Nethravathi, C.J.; Nitin, K.S. Soil Biodiversity and Arthropods: Role in Soil Fertility. In *Economic and Ecological Significance of Arthropods in Diversified Ecosystems*; Chakravarthy, A.K., Sridhara, S., Eds.; Springer: Singapore, 2016; pp. 17–51.

36. André, H.M.; Noti, M.-I.; Lebrun, P. The soil fauna: The other last biotic frontier. *Biodivers. Conserv.* **1994**, *3*, 45–56. [CrossRef]

37. van Straalen, N.M.; Verhoef, H.A. The Development of a Bioindicator System for Soil Acidity Based on Arthropod pH Preferences. *J. Appl. Ecol.* **1997**, *34*, 217–232. [CrossRef]

38. Lavelle, P.; Chauvel, A.; Fragoso, C. Faunal activity in acid soils. In *Plant-Soil Interactions at Low pH: Principles and Management. Developments in Plant and Soil Sciences*; Date, R.A., Grundon, N.J., Rayment, G.E., Probert, M.E., Eds.; Springer: Dordrecht, The Netherlands, 1995; Volume 64, pp. 201–211.

39. Andrés, P.; Moore, J.C.; Simpson, R.T.; Selby, G.; Cotrufo, F.; Denef, K.; Haddix, M.L.; Shaw, E.A.; de Tomasel, C.M.; Molowny-Horas, R.; et al. Soil food web stability in response to grazing in a semi-arid prairie: The importance of soil textural heterogeneity. *Soil Biol. Biochem.* **2016**, *97*, 131–143. [CrossRef]

40. Potapov, A.; Goncharov, A.; Semenina, E.; Korotkevich, A.; Tsurikov, S.; Rozanova, O.; Anichkin, A.; Zuev, A.; Samoylova, E.; Semenyuk, I.; et al. Arthropods in the subsoil: Abundance and vertical distribution as related to soil organic matter, microbial biomass and plant roots. *Eur. J. Soil Biol.* **2017**, *82*, 88–97. [CrossRef]

41. Shakir, M.M.; Ahmed, S. Seasonal abundance of soil arthropods in relation to meteorological and edaphic factors in the agroecosystems of Faisalabad, Punjab, Pakistan. *Int. J. Biometeorol.* **2015**, *59*, 605–616. [CrossRef]

42. Sapkota, T.B.; Mazzoncini, M.; Bàrberi, P.; Antichi, D.; Silvestri, N. Fifteen years of no till increase soil organic matter, microbial biomass and arthropod diversity in cover crop-based arable cropping systems. *Agron. Sustain. Dev.* **2012**, *32*, 853–863. [CrossRef]

43. Bedano, J.C.; Cantú, M.P.; Doucet, M.E. Soil springtails (Hexapoda: Collembola), symphylans and pauropods (Arthropoda: Myriapoda) under different management systems in agroecosystems of the subhumid Pampa (Argentina). *Eur. J. Soil Biol.* **2006**, *42*, 107–119. [CrossRef]

44. Chauvat, M.; Wolters, V.; Dauber, J. Response of collembolan communities to land-use change and grassland succession. *Ecography* **2007**, *30*, 183–192. [CrossRef]

45. Gkisakis, V.; Volakakis, N.; Kollaros, D.; Bàrberi, P.; Kabourakis, E.M. Soil arthropod community in the olive agroecosystem: Determined by environment and farming practices in different management systems and agroecological zones. *Agric. Ecosyst. Environ.* **2016**, *218*, 178–189. [CrossRef]

46. Hadjicharalampous, E.; Kalburtji, K.L.; Mamolos, A.P. Soil Arthropods (Coleoptera, Isopoda) in Organic and Conventional Agroecosystems. *Environ. Manag.* **2002**, *29*, 683–690. [CrossRef] [PubMed]

47. Gagnarli, E.; Goggioli, D.; Tarchi, F.; Guidi, S.; Nannelli, R.; Vignozzi, N.; Valboa, G.; Lottero, M.R.; Corino, L.; Simoni, S. Case study of microarthropod communities to assess soil quality in different managed vineyards. *Soil* **2015**, *1*, 527–536. [CrossRef]

48. Caprio, E.; Nervo, B.; Isaia, M.; Allegro, G.; Rolando, A. Organic versus conventional systems in viticulture: Comparative effects on spiders and carabids in vineyards and adjacent forests. *Agric. Syst.* **2015**, *136*, 61–69. [CrossRef]

49. Puig-Montserrat, X.; Stefanescu, C.; Torre, I.; Palet, J.; Fàbregas, E.; Dantart, J.; Arrizabalaga, A.; Flaquer, C. Effects of organic and conventional crop management on vineyard biodiversity. *Agric. Ecosyst. Environ.* **2017**, *243*, 19–26. [CrossRef]

50. Seniczak, A.; Seniczak, S.; García-Parra, I.; Ferragut, F.; Xamaní, P.; Graczyk, R.; Messeguer, E.; Laborda, R.; Rodrigo, E. Oribatid mites of conventional and organic vineyards in the Valencian Community, Spain. *Acarologia* **2018**, *58*, 119–133.

51. Costantini, E.A.C.; Agnelli, A.E.; Fabiani, A.; Gagnarli, E.; Mocali, S.; Priori, S.; Simoni, S.; Valboa, G. Short-term recovery of soil physical, chemical, micro- and mesobiological functions in a new vineyard under organic farming. *Soil* **2015**, *1*, 443–457. [CrossRef]

52. Döring, J.; Collins, C.; Frisch, M.; Kauer, R. Organic and Biodynamic Viticulture Affect Biodiversity and Vine and Wine Properties: A Systematic Quantitative Review. *Am. J. Enol. Vitic.* **2019**, *70*, 221–242. [CrossRef]

53. National Centers for Environmental Prediction; National Weather Service; NOAA; U.S. Department of Commerce. NCEP GDAS/FNL 0.25 Degree Global Tropospheric Analyses and Forecast Grids 2015. Research Data Archive at the National Center for Atmospheric Research, Computational and Information Systems Laboratory. Available online: https://doi.org/10.5065/D65Q4T4Z (accessed on 5 November 2019).

54. Powers, J.G.; Klemp, J.B.; Skamarock, W.C.; Davis, C.A.; Dudhia, J.; Gill, D.O.; Coen, J.L.; Gochis, D.J.; Ahmadov, R.; Peckham, S.E.; et al. The Weather Research and Forecasting Model: Overview, System Efforts, and Future Directions. *Bull. Am. Meteorol. Soc.* **2017**, *98*, 1717–1737. [CrossRef]

55. Ek, M.B.; Mitchell, K.E.; Lin, Y.; Rogers, E.; Grunmann, P.; Koren, V.; Gayno, G.; Tarpley, J.D. Implementation of Noah land surface model advances in the National Centers for Environmental Prediction operational mesoscale Eta model. *J. Geophys. Res. Atmos.* **2003**, *108*, 8851. [CrossRef]

56. Soil Science Division Staff. *Soil Survey Manual*; Ditzler, C., Scheffe, K., Monger, H.C., Eds.; USDA Handbook 18; Government Printing Office: Washington, DC, USA, 2017; p. 125.

57. Menta, C.; Conti, F.D.; Pinto, S.; Bodini, A. Soil Biological Quality index (QBS-ar): 15 years of application at global scale. *Ecol. Indic.* **2018**, *85*, 773–780. [CrossRef]

58. Kenkel, N.C.; Orloci, L. Applying Metric and Nonmetric Multidimensional Scaling to Ecological Studies: Some New Results. *Ecology* **1986**, *67*, 919–928. [CrossRef]

59. Clarke, K.R. Non-parametric multivariate analyses of changes in community structure. *Aust. J. Ecol.* **1993**, *18*, 117–143. [CrossRef]

60. Oksanen, J.; Blanchet, F.G.; Friendly, M.; Kindt, R.; Legendre, P.; McGlinn, D.; Minchin, P.R.; O'Hara, R.B.; Simpson, G.L.; Solymos, P.; et al. *Vegan: Community Ecology Package*, R Package Version 2.5; 2018. Available online: http://CRAN.Rproject.org/package=vegan (accessed on 19 May 2020).

61. R Core Team. *R: A Language and Environment for Statistical Computing. R Foundation for Statistical Computing*; R Core Team: Vienna, Austria, 2019. Available online: https://www.R-project.org/ (accessed on 19 May 2020).

62. Therneau, T.; Atkinson, B. *rpart: Recursive Partitioning and Regression Trees*, R Package Version 4.1-13; 2018. Available online: http://CRAN.Rproject.org/package=rpart (accessed on 19 May 2020).

63. Pérez Bote, J.; Romero, A. Epigeic soil arthropod abundance under different agricultural land uses. *Span. J. Agric. Res.* **2012**, *10*, 55–61. [CrossRef]

64. Zhu, G.; Luo, Y.; Xue, M.; Zhao, H.; Xia, N.; Wang, X. Effects of high-temperature stress and heat shock on two root maggots, Bradysia odoriphaga and Bradysia difformis (Diptera: Sciaridae). *J. Asia-Pac. Entomol.* **2018**, *21*, 106–114. [CrossRef]

65. Sjursen, H.; Holmstrup, M. Cold and drought stress in combination with pyrene exposure: Studies with Protaphorura armata (Collembola: Onychiuridae). *Ecotoxicol. Environ. Saf.* **2004**, *57*, 145–152. [CrossRef]

66. Heiniger, C.; Barot, S.; Ponge, J.-F.; Salmon, S.; Meriguet, J.; Carmignac, D.; Suillerot, M.; Dubs, F. Collembolan preferences for soil and microclimate in forest and pasture communities. *Soil Biol. Biochem.* **2015**, *86*, 181–192. [CrossRef]

67. Bertrand, M.; Lumaret, J.P. The role of Diplopoda litter grazing activity on recycling processes in a Mediterranean climate. *Vegetatio* **1992**, *99*, 289–297. [CrossRef]

68. Stašiovstasiov, S.; Stašiová, A.; Svitok, M.; Michalková, E.; Slobodník, B.; Lukáčikarbbh, I. Millipede (Diplopoda) communities in an arboretum: Influence of tree species and soil properties. *Biologia* **2012**, *67*, 945–952.

69. Badejo, M.A. Seasonal Abundance of Soil Mites (Acarina) in Two Contrasting Environments. *Biotropica* **1990**, *22*, 382–390. [CrossRef]

70. Byers, G.W.; Thornhill, R. Biology of the Mecoptera. *Annu. Rev. Entomol.* **1983**, *28*, 203–228. [CrossRef]

71. Varga, I. Structure and changes of macroinvertebrate community colonizing decomposing rhizome litter of common reed at Lake Fertő/Neusiedler See (Hungary). *Hydrobiologia* **2003**, *506*, 413–420. [CrossRef]

72. Weygoldt, P. *The Biology of Pseudoscorpions*; Harvard University Press: Cambridge, Massachusetts, 1969.

73. Bauer, T. Predation by a carabid beetle specialized for catching collembola. *Pedobiologia* **1982**, *24*, 169–179.

 *agronomy*

Article

The Effect of Cover Crops on the Biodiversity and Abundance of Ground-Dwelling Arthropods in a Mediterranean Pear Orchard

Luis de Pedro, Luis Gabriel Perera-Fernández, Elena López-Gallego, María Pérez-Marcos and Juan Antonio Sanchez *

Department of Crop Protection, Biological Control and Ecosystem Services, Instituto Murciano de Investigación y Desarrollo Agrario y Alimentario, C/Mayor s/n, La Alberca, 30150 Murcia, Spain; luis.depedro@carm.es (L.d.P.); lpererafernandez@gmail.com (L.G.P.-F.); elena.lopez5@carm.es (E.L.-G.); maria.perez28@carm.es (M.P.-M.)
* Correspondence: juana.sanchez23@carm.es; Tel.: +34-968-362-787

Received: 9 April 2020; Accepted: 16 April 2020; Published: 18 April 2020

Abstract: The intensification of agriculture has led to the reduction of the diversity of arthropods in agroecosystems, including that of ground-dwelling species. The aim of our work was to assess the effect of a sown cover crop on the diversity of ground-dwelling arthropods, including key predators for pest control in pear orchards. The trial was carried out in a pear orchard divided in three blocks; two treatments (cover-cropping and control) were implemented in each block. A seed mixture of 10 plant species was used in the plots with the sown cover. The densities of ground-dwelling arthropods were sampled using pitfall traps. The ground cover had a significant impact on the diversity and abundance of arthropods. The Shannon–Wiener diversity index was significantly higher for the cover than for the control plots. Several families of spiders (Linyphiidae, Lycosidae), beetles (Carabidae, Staphylinidae) and hymenopterans (Scelionidae) were significantly more abundant in the cover-sown plots. Ants and collembola had a significantly higher abundance in the control plots. Some of these groups arthropods (ants and spiders), are represented by species that may commute between ground and pear trees, having an impact on pest control. The use of cover crops is encouraged to enhance biodiversity in farmlands.

Keywords: ground-dwelling arthropods; pitfall traps; cover crops; ecosystem services; natural enemies; pear pests; biological control

---

1. Introduction

Biodiversity is currently experiencing one of the greatest known regressions since the beginning of life on Earth [1–3]. Under the current scenario, it is predicted that about 20% of all species will be lost in the next three decades [1,4]. Changes in land use and cover are currently considered the single-most acute factor threatening biodiversity worldwide, since native diversity depends on the structural and compositional diversity of habitats [5]. Among these changes, the conversion of natural ecosystems such as forests or grasslands to agriculture is considered to make a particularly high contribution [6]. Croplands and pastures are today one of the largest terrestrial biomes, occupying approximately 40% of the land surface on the planet [3]. In addition, the intensification of modern agriculture has resulted in the simplification of agricultural landscapes [7–9]. Habitat loss and fragmentation, combined with high inputs of pesticides, are nowadays considered the main causes of the worldwide loss of biodiversity [10–12].

Soil is one of the most species-rich habitats of terrestrial ecosystems [13–15]. According to diverse estimates, the soil fauna represents approximately 23% of all described organisms, with arthropods representing 85% of the species present in the soil fauna [16]. The arthropods that live on the soil surface ('ground-dwelling arthropods') also constitute an important part of the biodiversity of most terrestrial ecosystems [17,18]. The wide diversity of ground-dwelling arthropods includes several taxa that have a major presence in most of the surveys conducted in different ecosystems, such as Myriapoda, Collembola, Coleoptera (mainly carabids and staphylinids), Acari, Araneae and Formicidae [18–22]. Epigeic arthropods encompass a broad range of trophic guilds and ecological roles, thus influencing ecosystem function [17,18]. Many species of ground-dwelling arthropods do not spend their entire life on the soil surface, but commute between the ground and the aerial part of plants [13]. This is the case for many species of various groups of major predators, such as ants or spiders, which are ubiquitous in terrestrial ecosystems and essential to regulate the abundance of herbivores on plants [13,23]. Furthermore, many exclusively ground-dwelling arthropods may influence the population dynamics of aerial herbivores through cascading effects produced by "top-down" regulation processes, due to their interaction with commuting species [24–26]. For example, some carabids are known to feed on other predators of both the ground layer and the plant foliage, such as spiders, affecting their abundance via intraguild predation (IGP) and, consequently, the regulation of plant pest populations [27–29].

Common agricultural practices such as ploughing, the elimination of ruderal plants, and the use of fertilisers modify the conditions of soils and have a great impact on the diversity and abundance of epigeal arthropods, including many species that play a key role in the regulation of plant pests [30–33]. The relevance of biodiversity for the functioning of ecosystem processes together with the pivotal role that it plays in providing ecosystem services to humans makes it essential to plan conservation strategies to reverse the loss of species [5,34,35]. Biodiversity losses are associated with several key problems affecting the sustainability of farming systems, such as limited soil genesis and fertility, pollination, and pest control [36,37]. Because of the great extension of the Earth devoted to farming, conservation strategies aiming to increase the complexity of agricultural landscapes are expected to highly contribute to the maintenance of worldwide biodiversity and to the provision of ecosystem services [12]. Floral strips and cover crops are some of the agroecological practices used most frequently to enhance habitats of pollinating insects and natural enemies in environmentally degraded farmlands [3,34,38,39]. Green infrastructures are known to provide the missing habitat requirements for natural enemies (food resources, shelters, refuges, etc.), allowing them to overcome the disturbances derived from agricultural practices [40,41].

Fruit tree orchards may benefit from the adoption of agroecological practices, especially in simple landscapes. Orchards represent around 2% of the agricultural land utilised in the European Union (EU), with more than 3.4 million ha dedicated to fruit growing. Pears are one of the most important fruit crops in the EU. In 2018, more than 116,000 ha were devoted to pear production [42]. Therefore, increasing plant diversity in fruit tree orchards is expected to enhance biodiversity at a global scale, with a likely positive impact on ecosystem services such as pest control. A significantly higher abundance of natural enemies and improved pest control have been registered in fruit tree orchards with cover crops [39,43–45]. Pest control in pear orchards has traditionally relied on chemicals, but due to the restriction in the application of insecticides and the development of resistances, integrated pest management (IPM) has become the most-sustainable alternative [46–48]. Pear orchards with limited use of pesticides can be inhabited by a rich community of arthropods, which includes many natural enemies such as anthocorids, mirids, ants, and spiders that contribute to the regulation of the populations of herbivorous species [49–53]. In some parts of the Mediterranean area, ants (namely, *Lasius grandis* Forel, Hymenoptera: Formicidae) have been reported to be the key predator for the control of the pear psyllid [52,53]. This ant species spends the main part of its life cycle in the soil or on the soil surface; thus, its abundance and foraging activity may be greatly influenced by agricultural practices that modify soil conditions. Little information is available on the effect of cover crops on ground-dwelling invertebrates, especially the main groups of generalist predators [45].

Pear orchards are currently managed in a very intensive way, with the alleys between the lines of trees and the area surrounding the crop kept free from ruderal plants by ploughing or the regular use of herbicides. This way of farming is expected to have a high impact on the local diversity of arthropods, including some of the species that play a key role in the regulation of pests. Therefore, the aim of our study was to investigate how cover crops influence the diversity and abundance of ground-dwelling arthropods in a pear orchard. Predators that commute between the soil surface and the canopy of pear trees (e.g., ants and spiders) were of particular interest because of their likely impact on pest control.

2. Materials and Methods

2.1. Experimental Design

The present study was carried out in an organic pear orchard of approximately 5 ha (450 m-long, 110 m-wide) located near the locality of Jumilla (Murcia Province, 38°23′56″ N, 001°23′19″ W) in Southeastern Spain, during the spring of 2019. The effect of cover crops on the diversity and abundance of ground-dwelling arthropods was tested in a randomised block design experiment with three replicates of two treatments (i.e., cover crops and bare soil, Figure 1). The pear orchard had 26 lines of 540 trees each, with trees trained in trellises, the separation being 4 m between lines and 0.8 m between trees within lines. The orchard was divided in three blocks of approximately 1.6 hectares each. In each block, two plots, each 80 m-long and 20 m-wide (five lines of pear trees), separated by at least 4 lines of pear trees, were established. The two treatments were assigned randomly, one of the two plots of each of the three blocks being sown with a mixture of herbaceous plants, while the other maintained free from ruderal plants by periodical cuttings (every 2–3 weeks) and tillage. The mixture of seeds included the following herbaceous plants: *Borago officinalis* L., *Coriandrum sativum* L., *Calendula arvensis* L., *Calendula officinalis* L., *Diplotaxis erucoides* (L) DC., *Echium vulgare* L., *Hordeum vulgare* L., *Medicago sativa* L., *Phacelia tanacetifolia* Benth and *Vicia faba* L. These plant species were chosen with the aim of providing plentiful floral resources for beneficial arthropods and alternative prey/hosts for natural enemies and for improving soil fertility [54,55]. The pear trees were watered by above-ground drip irrigation twice a week; in addition, the sown plots were irrigated once a week by sprinklers to enhance the growth of the cover in the central part of the alleys between the lines of trees.

Figure 1. Example of the ground cover in the two treatments: the sown cover (**A**) and the control (**B**).

2.2. Sampling

The plots were monitored periodically in order to determine the effect of the sown cover on the structure of the community of ground-dwelling arthropods. The diversity and abundance of ground-dwelling arthropods were estimated using pitfall traps. Each trap consisted of a 500 mL plastic container (8 cm in diameter) partially filled with a mixture of water (94%), propylene glycol (5%) and soap (1%) and placed in the soil with its opening level with the soil surface. Three traps were set up in each of the two plots (i.e., cover and control) of each block; the traps were placed diagonally across the middle of the three central alleys of each plot. The traps were kept in the field for seven days, and then the specimens were collected and preserved in 70% alcohol for their identification. The samples were collected on 24 April, 13 May, 28 May and 11 June 2019. This period was chosen both because it is favourable for the development of the cover and because it is characterised by a high activity of insects and spiders [51,52]. The summer months in southern Spain are very arid, and plant covers dry out. The plots were sampled every two weeks because it was known from previous studies that the density of insects changes very little between two consecutive weeks [51,52].

The specimens collected were observed under a stereomicroscope and identified to the species level, whenever possible. When the identification to the species level was not possible, the specimens were assigned to morphospecies based on easily observable morphological characters [56]. The specimens were identified following the keys of Martínez et al. [57] for ants, Goulet and Huber [58] for other Hymenopterans, Nentwig et al. [59] for spiders and Salgado et al. [60] for beetles. The reference collection of voucher specimens is held by the IMIDA (Instituto Murciano de Investigación y Desarrollo Agrario y Alimentario).

The proportion of ground covered by vegetation for each plot was estimated by taking one high-resolution photograph, framing a 1 × 1 m plastic stick square, in each of the three alleys where pitfall traps were placed (i.e., 18 pictures per sampling date). The pictures were subdivided in 100 quadrants (10 cm × 10 cm), and the presence/absence of vegetation in each quadrant was scored. The GIMP v2.8.14 software (Free Software Foundation, Boston, MA, USA) was used for image processing.

2.3. Data Analysis

Generalised linear mixed-effects models (GLMM), run with the "lmer" function ("lme4" package) for normally distributed data, were used to compare the proportion of ground cover between the plots with and without the sown cover [61]; block and date of sampling were introduced in the models as random factors.

To test for the effect of ground cover, only the species that live on the ground or that spend part of their lives on the ground were considered in this study. The following taxa were included: Collembola, four families of Coleoptera (Anthicidae, Tenebrionidae, Carabidae and Staphylinidae), four families of Araneae (Gnaphosidae, Zodariidae, Lycosidae and Linyphiidae) and two families of Hymenoptera (Formicidae and Scelionidae).

The richness of species/morphospecies and the Shannon–Wiener diversity index were used to test for the effect of ground cover on the diversity of ground-dwelling arthropods. The effect of sown cover on the number of species/morphospecies of ants, spiders and beetles was estimated using GLMM run with the function "glmmPQL" (library "MASS") set for normal distributed data, i.e., family = gaussian (link = "identity"), in R [62]. Block and date of sampling were introduced in the models as random factors. The same procedure was used to estimate the effect of the sown cover on the Shannon–Wiener diversity index. The diversity index was calculated for each sampling date using the total number of captures of each of the species/morphospecies of the above-mentioned families, with the "diversity" function in the "vegan" package in R [62]. The χ^2- and *p*-values were obtained using the "Anova" function in the R "car" package [62].

The assemblages of ground-dwelling arthropods were compared between the plots with cover and the controls by PERMANOVA, using the function "adonis", the Euclidean distances being calculated with the "vegdist" function; these two functions are available in the "vegan" package in R [62]. The number of specimens (i.e., the sum of the three pitfall traps in each plot) of the abovementioned families of ground-dwelling arthropods collected on each sampling date were introduced in the models as dependent variables. Redundancy analyses (RDA) were applied to find out how samples clustered in relation to the presence/absence of the sown cover. The function "rda" in the "vegan" package was used to perform RDA on the number of ground-dwelling arthropods of the different families collected in the plots with the sown cover and the control plots on each sampling date; the captures of the three pitfall traps for each plot and sampling date were summed for every family of arthropods.

To determine the contribution of the abundance of every family of ground-dwelling arthropods —as a dependent variable—to the differences between the plots with cover and the controls—type of cover as fixed factor–GLMM were used. The "lmer" function ("lme4" package) was used to perform these analyses [61]; block and date of sampling were introduced in the models as random factors. For all the families, the numbers of captures were transformed by the natural logarithm of $(x + 1)$ to correct the deviation of the data from normality. The χ^2- and p-values were obtained as explained above.

3. Results

3.1. Ground Cover and Diversity of Ground-Dwelling Arthropods

The proportion of ground covered with vegetation was significantly higher in the plots sown with the mixture of seeds than in the control plots ($\chi^2 = 61.38$, df $= 1$, $p < 0.001$). The ground of the sown cover plots was almost entirely covered with vegetation during the whole sampling period, while in the control plots, the proportion of cover was very low on the first sampling date (0.143 ± 0.029), increasing to 0.718 ± 0.067 at the end of the experiment.

Along this study, a total of 25,139 arthropods were captured in the pitfall traps, with Collembola representing most of the captures (79.7%) (Supplementary Material, Table S1). Excluding Collembola, the most abundant arthropods collected in the pitfall traps were ants (76.0%), followed by Coleoptera (13.8%), spiders (8.1%) and scelionids (2.1%).

The richness of ground-dwelling species in cover and control plots varied in the different orders of arthropods (Figure 2A). Hymenopterans were mainly represented by ant species (Supplementary Material, Table S1), and their richness was significantly lower in the plots with cover than in the controls ($\chi^2 = 3.91$, df $= 1$, $p = 0.048$). The number of species of hymenopterans collected in the pitfall traps experienced little variation, the highest values being registered at the end of the experiment, in both the cover (3.7 ± 0.3, mean $\pm$ SE) and the control plots (5.0 ± 0.0). In contrast, the richness of spiders was significantly higher with a sown cover than in the control plots ($\chi^2 = 17.79$, df $= 1$, $p < 0.001$). In the cover plots, the number of species of spiders was the lowest (2.7 ± 0.7) in the first week of sampling and reached its maximum (7.3 ± 0.9) at the beginning of May. In the control plots, the lowest (1.3 ± 0.9) and highest (5.0 ± 0.6) numbers of spiders were registered at the end of April and May, respectively. In the same way, the richness of beetles was also significantly higher in the grounds with a sown cover than in the control plots ($\chi^2 = 16.65$, df $= 1$, $p < 0.001$). The trend in the number of species of beetles was very similar to that of spiders (Figure 2A). In the cover plots, the lowest (3.3 ± 0.7) and highest (7.0 ± 1.2) values were registered at the end of April and May, respectively. In the control plots, the numbers of species of beetles increased progressively from the beginning until the end of the study, ranging between 3.3 ± 0.7 and 4.3 ± 0.3. No distinction among species/morphospecies was made in springtails.

The Shannon–Wiener diversity index of ground-dwelling arthropods was significantly higher in the plots with a sown cover than in the control plots ($\chi^2 = 25.52$, df $= 1$, $p < 0.001$) (Figure 2B). The plots with cover showed a progressive increase in the Shannon–Wiener diversity index throughout the period of study, reaching the highest value at the end of the study in June (2.04 ± 0.13). In the control plots, the diversity index varied little among the sampling dates, reaching its lowest value at the end of May (0.77 ± 0.16); thereafter, it increased until June (1.20 ± 0.16).

3.2. Structure of the Assemblages of Ground-Dwelling Arthropods in Pear Orchards

The plots with and without the sown cover differed in their assemblages of ground-dwelling arthropods (PERMANOVA, F $= 2.44$, df $= 1$, 22, $p = 0.030$). In the RDA analysis, practically all the samples from the plots with the sown cover clustered on the positive side of the first component of RDA, while the samples from the control plots grouped on the negative side (Figure 3). The first constrained axis (RDA1) explained 16.5% of the variance in relation to cover (F $= 4.35$, df. 1, 22, $p < 0.001$). Carabidae, Linyphiidae, Staphylinidae and Lycosidae were the families of arthropods with the highest correlation in relation to cover. In contrast, Collembola and Formicidae were highly correlated with plots without sown cover (Figure 3).

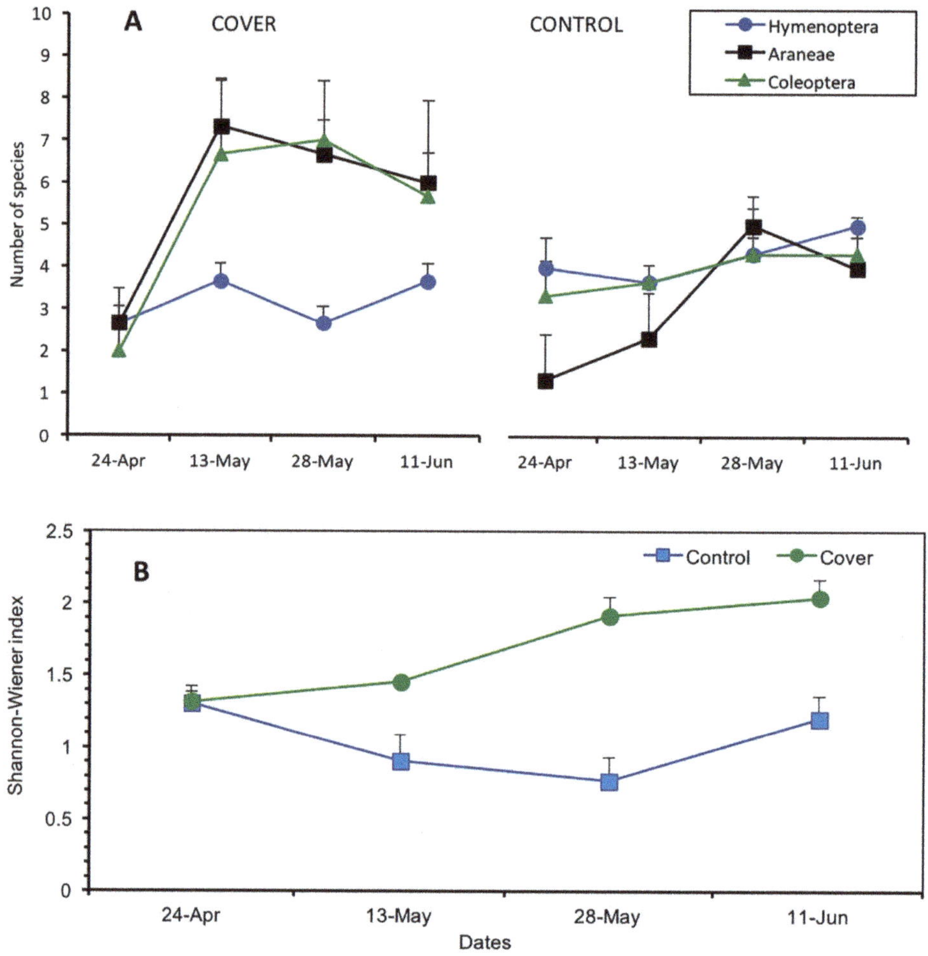

Figure 2. (**A**) Richness of hymenopterans, spiders and beetles; (**B**) Shannon–Wiener index in the plots with a sown cover and in the control plots (mean ± SE).

The abundance of most of the families of arthropods collected in the pitfall traps, with the exception of some Araneae (i.e., Gnaphosidae and Zodariidae) and Coleoptera (i.e., Anthicidae and Tenebrionidae), differed significantly between the plots with cover and the controls (Table 1). Ants were represented by polyphagous species that may potentially commute between the ground and the aerial part of pear trees. *L. grandis* was the most abundant ant species (61.7%), followed by *Tetramorium* spp. (28.9%) and other minor species (<5%) such as *Formica* spp., *Cataglyphis* spp., *Cardiocondyla* spp. and *Solenopsis* spp. (Supplementary Material, Table S1). Ant numbers peaked in the plots with a cover in mid-May (176.3 ± 26.7, mean of the total number of individuals collected per plot ± SE) and in the control plots at the end of May (316.0 ± 36.1) (Figure 4).

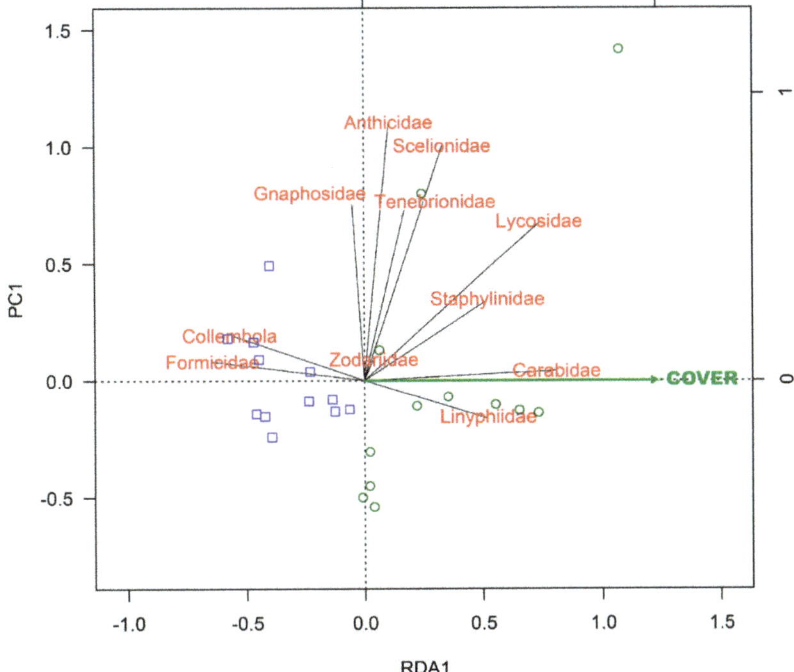

Figure 3. Plot of the first constrained ordination axis (RDA1) versus the first unconstrained axis (PC1) on the redundancy analyses (RDA) of samples of ground-dwelling arthropods collected in plots with (green circles) and without (blue squares) a sown cover.

Table 1. Coefficients and statistics of the generalised linear mixed effects models (GLMM) to test for the effect of the cover crop on the abundance of the main groups of ground-dwelling arthropods. χ^2 = Chi square; df = degrees of freedom.

Order	Family	Coefficient	χ^2	df	p
Hymenoptera	Formicidae	−0.664	24.032	1	<0.001
	Scelionidae	0.710	5.161	1	0.023
Araneae	Gnaphosidae	−0.227	0.720	1	0.396
	Linyphiidae	0.674	9.705	1	0.002
	Lycosidae	1.799	67.751	1	<0.001
	Zodariidae	−0.010	0.002	1	0.970
Coleoptera	Anthicidae	0.144	0.336	1	0.562
	Carabidae	2.058	43.180	1	<0.001
	Staphylinidae	0.806	6.008	1	0.014
	Tenebrionidae	0.097	0.071	1	0.790
Collembola	-	−0.650	10.063	1	0.002

In the case of spiders, most of the families collected in the pitfall traps (i.e., Gnaphosidae, Lycosidae and Zodariidae) forage on the ground, while Linyphiidae are also found on the canopy. The most abundant family of spiders was Lycosidae (45.6%), followed by Gnaphosidae (27.0%), Zodariidae (15.5%) and Linyphiidae (11.8%). The highest number of spiders collected belonged to *Pardosa* spp. (39.4% of the captures), *Micaria* spp. (19.3%) and *Zodarion* spp. (15.5%) (Supplementary Material, Table S1). The abundance of Lycosidae and Linyphiidae was significantly higher in the cover plots than in the control plots (Table 1). In the plots with a sown cover, the abundances of these two families gradually increased until the end of May, when lycosids reached the highest values recorded among the spiders

(26.0 ± 4.0); the linyphiids reached a much lower peak (7.0 ± 2.7) (Figure 4). In the control plots, the abundances of lycosids and linyphiids were very low. In contrast, the abundances of Gnaphosidae and Zodariidae did not differ significantly between the two treatments (Table 1), with very similar numbers of specimens captured in both types of plot along the study (Supplementary Material, Table S1). These two families peaked at different times: the zodariids at the end of May (Control: 5.7 ± 0.7; Cover: 5.7 ± 1.3), and the gnaphosids at the beginning of June (Control: 12.0 ± 1.2; Cover: 11.3 ± 4.1) (Figure 4).

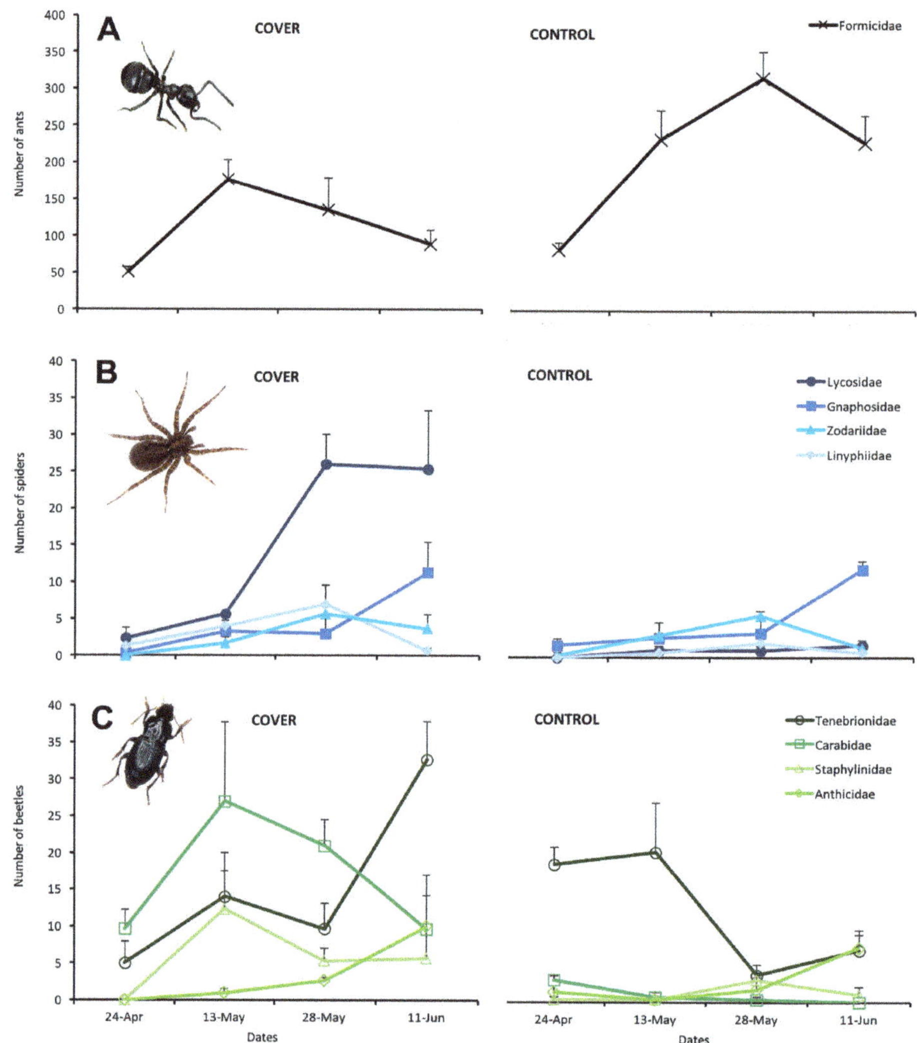

Figure 4. Number of the different families of ants (**A**), spiders (**B**) and beetles (**C**) (mean ± SE) collected in pitfall traps in the plots with a sown cover and in the control plots.

Beetles were represented by families with different feeding habits. Phytophagous species of the families Tenebrionidae (47.4%) and Carabidae (namely, *Harpalus* spp., 30.4%) represented most of the Coleoptera collected in pitfall traps; polyphagous species, such as staphylinids (11.9%) and anthicids (10.2%), were less represented (Supplementary Material, Table S1). The four families of beetles showed different trends in their abundances along the sampling period in relation to the type of cover (Figure 4). Tenebrionids and anthicids did not show significant differences between the plots with a cover and the controls (Table 1). In contrast, carabids, represented only by the species *Harpalus*, and staphylinids were much more numerous in the plots with a sown cover (Table 1). The abundance of tenebrionids increased in the cover plots to reach a peak in June (32.7 ± 5.2 individuals), while in the controls, the lowest abundances were registered on the last two sampling dates. Carabids and staphylinids peaked in mid-May in the plots with a sown cover (Carabidae: 27.0 ± 10.8; Staphylinidae: 12.3 ± 5.2), their abundances decreasing thereafter. In the controls, these families were scarcer, with carabids (3.0 ± 0.6) peaking at the end of April, and staphylinids (3.0 ± 1.0) at the end of May. The numbers of anthicids, despite being generally low, gradually increased along the sampling period, peaking on the last sampling date in the two treatments (Control: 7.3 ± 2.4; Cover: 10.0 ± 4.1).

Scelionids were collected only occasionally, but they are relevant for being egg parasitoids of arthropods. They were mainly represented by the genus *Baeus* Haliday (93.4% of the captures). These hymenopterans were significantly more numerous in the cover plots than in the control plots (Table 1). The abundance of scelionids was very low in the first three samplings and increased considerably at the beginning of June in the cover plots (32.0 ± 19.3), relative to the control plots (1.7 ± 1.2). Finally, springtails showed significantly higher abundances in the control plots than in those with a sown cover (Table 1). Springtails gradually increased in number in the control plots, peaking in June at 2343.3 ± 378.9 individuals; in contrast, in the cover plots they peaked in mid-May (1068.3 ± 149.5 individuals), with their abundances decreasing thereafter.

4. Discussion

Agroecological practices such as the implementation of cover crops are known to contribute to the maintenance of local biodiversity in farming systems [3,12,34]. The results of the present work indicate that a rich cover of vegetation increases the biodiversity of ground-dwelling arthropods in pear orchards. The Shannon–Wiener diversity index was significantly higher in the presence than in the absence of a sown cover. In addition, the richness of spiders and beetles was significantly higher in the plots with a cover. A mix of herbaceous plants similar to the one used in the present work was reported to produce an increase in the abundance and diversity of wild bees in areas of intensive agriculture [55]. Considering the growing interest in green infrastructures for the conservation of biodiversity in agricultural lands, relatively little information on the impact of cover crops on the diversity of ground-dwelling arthropods is available. Sommaggio et al. [45] found a significant higher activity and density of isopods, staphylinids, carabids and grillids in the soil surface of a vineyard with several types of cover crop, relative to the control, which was exposed to periodical tillage; however, only a faba bean cover had a significantly higher number of species than the control. Surprisingly, no significant differences were found between any of the treatments and the control for the Shannon–Wiener index, with the exception of a buckwheat cover that registered lower values that the control. In contrast, Cárdenas et al. [63] found no significant differences in spider diversity between ground with cover and that where the vegetation had been removed. Rieux et al. [64] reported a higher diversity index for arthropods on sown cover than on bare ground and natural vegetation cover in French pear orchards. However, it has to be taken into account that, because sampling was carried out using sweeping nets, these indexes represent the diversity of arthropods living on plants rather than of those living on the ground.

Most of the main groups of ground-dwelling arthropods collected were significantly affected by the presence of a sown cover. Among them, only springtails and ants showed lower abundances on the ground with a sown cover than on the ground without a cover; additionally, the ant richness was lower on the ground with a cover. This is in contrast with previous studies reporting higher ant abundances under cover-cropping management [65–68]; however, it should be noted that most of these studies compared soils with cover crops with recently tilled soils, and intense tillage is known to have a detrimental effect on ant abundances [69]. Regarding springtails, our results are in agreement with those of Buchholz et al. [70], who stated that the abundance and diversity of surface-dwelling springtails were diminished by the greater plant biomass provided by covers. Beetles (i.e., carabids and staphylinids) and spiders (i.e., lycosids and linyphiids) were more numerous on the ground with a sown cover. The beneficial effects of covers on carabids and staphylinids have been extensively reported [45,71,72], while in the case of spiders the results are more variable. Several studies have reported an increase in the abundance of spiders on the ground in orchards with a vegetation cover [73–75], while in other studies, a non-significant effect in comparison to bare ground was registered [45,63]. The scelionids were another group of insects that benefited from the sown cover. Other authors have also reported increased abundances of scelionids on grounds with cover, in different types of orchards [76,77].

Cover crops may affect ground-dwelling arthropods in several ways. For instance, by creating physical barriers that hamper their movement on the ground surface and/or by increasing the availability of niches in habitats [70,78,79]. In the present work, these two factors could explain the decline of springtails observed in the plots with a sown cover. Buchholz et al. [70] argued that plant covers not only hinder the rapid movement of springtails, increasing the risk of them falling prey, but also benefit the establishment of predators. In our case, lycosids, that benefited from the sown covers, are known to prey on springtails [80,81]. In relation to ants, very few species of ant predators—restricted to a few families of spiders—have been reported in agroecosystems [82,83]. Therefore, in the present work, the lower abundance of ants (namely, *L. grandis*) registered on the ground with cover was more likely due to physical interference and/or to interactions with other species. For instance, several herbaceous plants included in the cover host ant-mutualistic aphids that may divert the attention of ants to these plants [44,84,85]. In the case of spiders, a significant increase in the number of lycosids and linyphiids was registered on the ground with more vegetation. These two families have been reported to benefit from the structural complexity and hideouts provided by herbaceous plants [78,79]. Moreover, these plants may increase the availability of phytophagous and saprophagous prey, which constitute a great part of the diet of these spiders [86,87]. The abundance of spiders could explain the higher number of scelionids in the plots with a sown cover crop. This hymenopteran family was mostly represented by the genus *Baeus*, an obligate parasitoid of spider eggs known to target egg sacs of *Pardosa* wolf spiders [88]. In the case of the two main families of ground-dwelling coleopterans found in the present work, carabids and staphylinids, the factors that may have contributed to their increase on the ground with a sown cover are not easy to determine. Most previous studies focused on predatory species and argued that an improved physical structure of microhabitats, higher alternative food availability, reduced competition and/or an increase in the prey population could be the main explanations for higher densities of these beetles in cover crops [45,71]. In this study, all the carabids collected belonged to the genus *Harpalus* Latreille, which includes mostly phytophagous species [45]. In this regard, Shearin et al. [33] observed a beneficial effect of cover crops on the abundance of the species *Harpalus rufipes* De Geer (Coleoptera: Carabidae), suggesting higher seed availability as the main factor behind this trend. Ground-dwelling arthropods are also influenced by the variation in microclimatic conditions due to cover crops [89,90]. Vegetation gives shelter to ground-dwelling arthropods against extreme temperatures and provides higher environmental humidity. The abundances of carabids and spiders have frequently been found to be positively correlated with soil moisture [91,92]. In particular, the higher recaptures of *H. rufipes* in a cover crop, in comparison to fallow treatments, were attributed to higher humidity and lower temperature [33]. In the present study, the increase in humidity produced by the greater vegetation cover and the extra watering of the cover crop may have also benefited some arthropods, such as carabids and spiders.

This and earlier studies have demonstrated that vegetation covers allow the existence of a more abundant and diverse arthropofauna in crops [67,93,94]. Cover crops may increase the availability of resources (e.g., pollen, nectar, alternative host and prey species, shelter) to support a rich community of natural enemies that may eventually move to adjacent crop plants and exert a beneficial effect [34,95,96]. Evidence of generalist predators, such as spiders, commuting between a legume cover crop and the canopy of pear trees has been found using immunomarkers [97]. Although the relationship between biodiversity and ecosystem functioning is controversial [12], increasing the diversity of vegetation in crops has frequently been reported to enhance ecosystem services such as pest control [93,98]. Several studies have provided evidence of plant covers enhancing the abundance of natural enemies and pest control in fruit tree orchards [39,44,45,99,100]. In the case of pear trees, there is some evidence of a positive effect of ground covers on beneficial fauna [64,97,101]. In the present study, some of the ground-dwelling arthropods influenced by the plant cover, namely, ants and spiders, are key species for the assemblage of arthropods in pear orchards in the Mediterranean area [51–53]. Therefore, increasing the herbaceous vegetation in pear orchards is expected to have an impact on the population dynamics of the species in the canopy of the trees. However, the outcome of the interaction among species is difficult to predict. Ants are known to establish antagonistic–mutualistic interactions with psyllids, being the key species for the control of pear psyllids in some parts of the Mediterranean area [52,53]; thus, the foreseen change in the foraging pattern of ants due to increased vegetation may have either a positive or a negative effect on the control of pear psyllids. The effect of cover crops on spiders as biological control agents is expected to be lower than that on ants, especially because they are much less numerous than ants [51] and because, of the families found in the canopy of pear trees (J. A. Sanchez, non-published data), only the Linyphiidae were found to be influenced by the cover. Other spiders not affected by the cover, such as the genus *Zodarion* Walckenaer, have been described as specialist ant predators that prey on medium-sized ants, such as *Lasius* spp. [83,102]. However, the impact of *Zodarion* spp. on ants is expected to be low because of their low abundance.

In the present work, it was found that cover crops had a significant effect on the diversity of ground-dwelling arthropods, including some key predators for the control of pests in pear orchards, such as ants and spiders. This work outlines how agroecological practices may contribute to the maintenance of local biodiversity and the importance of including farmlands in the plans for the conservation of the species. The impact of cover crops in terms of pest control is uncertain; therefore, more work is needed to determine how cover crops affect the population dynamics of pests and predators in the aerial part of pear trees, as well as how the interactions among species on the ground influence population dynamics in the canopy. Although this work was carried out only during one year and over a short period of time, it provides evidence that plant covers influence the diversity of ground-dwelling arthropods. Samplings over a more extensive period will reveal the impact of cover crops under different environmental conditions and on other groups of arthropods that had little representation in this study.

Supplementary Materials: The following are available online at http://www.mdpi.com/2073-4395/10/4/580/s1, Table S1. List of taxa collected in pitfall traps in the plots with a sown cover and in the control plots.

Author Contributions: Conceptualization and Methodology, L.d.P., L.G.P.-F. and J.A.S.; Investigation, L.d.P., L.G.P.-F., E.L.-G., M.P.-M. and J.A.S.; Analyses of data, J.A.S.; Writing—original draft preparation, L.d.P., L.G.P-F. and J.A.S.; Writing—review and editing, J.A.S. All authors have read and agreed to the published version of the manuscript.

Funding: This research was funded by the project FEDER 1420-19 (European Regional Development Fund).

Acknowledgments: We thank the grower Antonio García (La Tierrica Bio) for allowing us access to his orchards to carry out this work. We also thank Celia Sánchez Marín for technical assistance.

Conflicts of Interest: The authors declare no conflict of interest. The funders had no role in the design of the study; in the collection, analyses, or interpretation of data; in the writing of the manuscript, or in the decision to publish the results.

References

1. Singh, J.S. The biodiversity crisis: A multifaceted review. *Curr. Sci.* **2002**, *82*, 638–647.
2. Rosengrant, M.; Cai, X. *World Water and Food to 2025*; International Food Policy Research Institute: Washington, DC, USA, 2002; ISBN 0896296466.
3. Foley, J.A.; DeFries, R.; Asner, G.P.; Barford, C.; Bonan, G.; Carpenter, S.R.; Chapin, F.S.; Coe, M.T.; Daily, G.C.; Gibbs, H.K.; et al. Global consequences of land use. *Science* **2005**, *309*, 570–574. [CrossRef]

4. Myers, N. Biodiversity and the Precautionary Principle. *Ambio* **1993**, *22*, 74–79.

5. Hunter, M.; Gibbs, J. *Fundamentals of Conservation Biology*; Wiley: Hoboken, NJ, USA, 2007.

6. Baldwin, R.F. Identifying Keystone Threats to Biological Diversity. In *Landscape-Scale Conservation Planning*; Trombulak, S.C., Baldwin, R., Eds.; Springer: Berlin, Germany, 2010; pp. 17–32. ISBN 9789048195749.

7. Robinson, R.A.; Sutherland, W.J. Post-war changes in arable farming and biodiversity in Great Britain. *J. Appl. Ecol.* **2002**, *39*, 157–176. [CrossRef]

8. Bianchi, F.J.J.A.; Booij, C.J.H.; Tscharntke, T. Sustainable pest regulation in agricultural landscapes: A review on landscape composition, biodiversity and natural pest control. *Proc. R. Soc. B Biol. Sci.* **2006**, *273*, 1715–1727. [CrossRef]

9. Bommarco, R.; Kleijn, D.; Potts, S.G. Ecological intensification: Harnessing ecosystem services for food security. *Trends Ecol. Evol.* **2013**, *28*, 230–238. [CrossRef]

10. Matson, P.A.; Parton, W.J.; Power, A.G.; Swift, M.J. Agricultural Intensification and Ecosystem Properties. *Science* **1997**, *277*, 504–508. [CrossRef]

11. Tilman, D.; Fargione, J.; Wolff, B.; D'Antonio, C.; Dobson, A.; Howarth, R.; Schindler, D.; Schlesinger, W.H.; Simberloff, D.; Swackhamer, D. Forecasting Agriculturally Driven Global Environmental Change. *Science* **2001**, *292*, 281–284. [CrossRef]

12. Tscharntke, T.; Klein, A.M.; Kruess, A.; Steffan-Dewenter, I.; Thies, C. Landscape perspectives on agricultural intensification and biodiversity—Ecosystem service management. *Ecol. Lett.* **2005**, *8*, 857–874. [CrossRef]

13. Wolters, V. Biodiversity of soil animals and its function. *Eur. J. Soil Biol.* **2001**, *37*, 221–227. [CrossRef]

14. Decaëns, T.; Jiménez, J.J.; Gioia, C.; Measey, G.J.; Lavelle, P. The values of soil animals for conservation biology. *Eur. J. Soil Biol.* **2006**, *42*, S23–S38. [CrossRef]

15. Briones, M.J.I. Soil fauna and soil functions: A jigsaw puzzle. *Front. Environ. Sci.* **2014**, *2*, 1–22. [CrossRef]

16. Culliney, T.W. Role of arthropods in maintaining soil fertility. *Agriculture* **2013**, *3*, 629–659. [CrossRef]

17. Abbott, I.; Parker, C.A.; Sills, I.D. Changes in the Abundance of Large Soil Animals and Physical Properties of Soils Following Cultivation. *Aust. J. Soil Res.* **1979**, *17*, 343–353. [CrossRef]

18. Simão, F.C.P.; Carretero, M.A.; do Amaral, M.J.A.; Soares, A.M.V.d.M.; Mateos, E. Composition and seasonal variation of epigeic arthropods in field margins of NW Portugal. *Turk. J. Zool.* **2015**, *39*, 404–411.

19. Yi, H.; Moldenke, A. Response of Ground-Dwelling Arthropods to Different Thinning Intensities in Young Douglas Fir Forests of Western Oregon. *Environ. Entomol.* **2005**, *34*, 1071–1080. [CrossRef]

20. Torres, J.B.; Ruberson, J.R. Abundance and diversity of ground-dwelling arthropods of pest management importance in commercial Bt and non-Bt cotton fields. *Ann. Appl. Biol.* **2007**, *150*, 27–39. [CrossRef]

21. Meyer, W.M.; Eble, J.A.; Franklin, K.; McManus, R.B.; Brantley, S.L.; Henkel, J.; Marek, P.E.; Hall, W.E.; Olson, C.A.; McInroy, R.; et al. Ground-Dwelling Arthropod Communities of a Sky Island Mountain Range in Southeastern Arizona, USA: Obtaining a Baseline for Assessing the Effects of Climate Change. *PLoS ONE* **2015**, *10*, e0135210. [CrossRef]

22. Jabbour, R.; Pisani-Gareau, T.; Smith, R.G.; Mullen, C.; Barbercheck, M. Cover crop and tillage intensities alter ground-dwelling arthropod communities during the transition to organic production. *Renew. Agric. Food Syst.* **2016**, *31*, 361–374. [CrossRef]

23. Stefani, V.; Pires, T.L.; Torezan-Silingardi, H.M.; Del-Claro, K.; Ballhorn, D. Beneficial effects of ants and spiders on the reproductive value of *Eriotheca gracilipes* (Malvaceae) in a tropical savanna. *PLoS ONE* **2015**, *10*, e0131843. [CrossRef]

24. Mills, N.J. Factors influencing top-down control of insect pest populations in biological control systems. *Basic Appl. Ecol.* **2001**, *332*, 323–332. [CrossRef]

25. Aguilar-Fenollosa, E.; Ibáñez-Gual, M.V.; Pascual-Ruiz, S.; Hurtado, M.; Jacas, J.A. Effect of ground-cover management on spider mites and their phytoseiid natural enemies in clementine mandarin orchards (II): Top-down regulation mechanisms. *Biol. Control* **2011**, *59*, 171–179. [CrossRef]

26. Ratnadass, A.; Fernandes, P.; Avelino, J.; Habib, R. Plant species diversity for sustainable management of crop pests and diseases in agroecosystems: A review. *Agron. Sustain. Dev.* **2012**, *32*, 273–303. [CrossRef]

27. Snyder, W.E.; Wise, D.H. Predator Interference and the Establishment of Generalist Predator Populations for Biocontrol. *Biol. Control* **1999**, *15*, 283–292. [CrossRef]

28. Lang, A. Intraguild interference and biocontrol effects of generalist predators in a winter wheat field. *Oecologia* **2003**, *134*, 144–153. [CrossRef]

29. Davey, J.S.; Vaughan, I.P.; Andrew King, R.; Bell, J.R.; Bohan, D.A.; Bruford, M.W.; Holland, J.M.; Symondson, W.O.C. Intraguild predation in winter wheat: Prey choice by a common epigeal carabid consuming spiders. *J. Appl. Ecol.* **2013**, *50*, 271–279. [CrossRef]

30. Lundgren, J.G.; Shaw, J.T.; Zaborski, E.R.; Eastman, C.E. The influence of organic transition systems on beneficial ground-dwelling arthropods and predation of insects and weed seeds. *Renew. Agric. Food Syst.* **2006**, *21*, 227–237. [CrossRef]

31. Pullaro, T.C.; Marino, P.C.; Jackson, D.M.; Harrison, H.F.; Keinath, A.P. Effects of killed cover crop mulch on weeds, weed seeds, and herbivores. *Agric. Ecosyst. Environ.* **2006**, *115*, 97–104. [CrossRef]

32. De Aquino, A.M.; Ferreira da Silva, R.; Mercante, F.M.; Fernandes Correia, M.E.; de Fátima Guimarães, M.; Lavelle, P. Invertebrate soil macrofauna under different ground cover plants in the no-till system in the Cerrado. *Eur. J. Soil Biol.* **2008**, *44*, 191–197. [CrossRef]

33. Shearin, A.F.; Chris Reberg-Horton, S.; Gallandt, E.R. Cover Crop Effects on the Activity-Density of the Weed Seed Predator *Harpalus rufipes* (Coleoptera: Carabidae). *Weed Sci.* **2008**, *56*, 442–450. [CrossRef]

34. Landis, D.A.; Wratten, S.D.; Gurr, G.M. Habitat Management to Conserve Natural Enemies of Arthropod Pests in Agriculture. *Annu. Rev. Entomol.* **2000**, *45*, 175–201. [CrossRef]

35. Foley, J.A.; Ramankutty, N.; Brauman, K.A.; Cassidy, E.S.; Gerber, J.S.; Johnston, M.; Mueller, N.D.; O'Connell, C.; Ray, D.K.; West, P.C.; et al. Solutions for a cultivated planet. *Nature* **2011**, *478*, 337–342. [CrossRef] [PubMed]

36. Hooper, D.U.; Chapin, F.S.; Ewel, J.J.; Hector, A.; Inchausti, P.; Lavorel, S.; Lawton, J.H.; Lodge, D.M.; Loreau, M.; Naeem, S.; et al. Effects of biodiversity on ecosystem functioning: A consensus of current knowledge. *Ecol. Monogr.* **2005**, *75*, 3–35. [CrossRef]

37. Overpeck, J.; Garfin, G.; Jardine, A.; Busch, D.E.; Cayan, D.; Dettinger, M.; Fleishman, E.; Gershunov, A.; MacDonald, G.; Redmond, K.T.; et al. *Summary for Decision Makers*; Island Press: Washington DC, USA, 2013; ISBN 9781610914840.

38. Thies, C.; Tscharntke, T. Landscape structure and biological control in agroecosystems. *Science* **1999**, *285*, 893–895. [CrossRef] [PubMed]

39. Silva, E.B.; Franco, J.C.; Vasconcelos, T.; Branco, M. Effect of ground cover vegetation on the abundance and diversity of beneficial arthropods in citrus orchards. *Bull. Entomol. Res.* **2010**, *100*, 489–499. [CrossRef]

40. Bugg, R.L.; Pickett, C.H. Introduction: Enhancing biological control-habitat management to promote natural enemies of agricultural pests. In *Enhancing Biological Control*; Pickett, C.H., Bugg, R.L., Eds.; University of California Press: Berkeley, CA, USA, 1998; pp. 1–23.

41. Jonsson, M.; Wratten, S.D.; Landis, D.A.; Gurr, G.M. Recent advances in conservation biological control of arthropods by arthropods. *Biol. Control* **2008**, *45*, 172–175. [CrossRef]

42. Eurostat. Available online: https://ec.europa.eu/eurostat/data/database (accessed on 20 February 2020).

43. Aguilar-Fenollosa, E.; Ibáñez-Gual, M.V.; Pascual-Ruiz, S.; Hurtado, M.; Jacas, J.A. Effect of ground-cover management on spider mites and their phytoseiid natural enemies in clementine mandarin orchards (I): Bottom-up regulation mechanisms. *Biol. Control* **2011**, *59*, 158–170. [CrossRef]

44. Gómez-Marco, F.; Urbaneja, A.; Tena, A. A sown grass cover enriched with wild forb plants improves the biological control of aphids in citrus. *Basic Appl. Ecol.* **2016**, *17*, 210–219. [CrossRef]

45. Sommaggio, D.; Peretti, E.; Burgio, G. The effect of cover plants management on soil invertebrate fauna in vineyard in Northern Italy. *BioControl* **2018**, *63*, 795–806. [CrossRef]

46. Berrada, S.; Fournier, D.; Cuany, A.; Nguyen, T. Identification of Resistance Mechanisms in a Selected Laboratory Strain of *Cacopsylla pyri* (Homoptera: Psyllidae): Altered Acetylcholinesterase and Detoxifying Oxidases. *Pestic. Biochem. Physiol.* **1994**, *48*, 41–47. [CrossRef]

47. Bues, R.; Boudinhon, L.; Toubon, J. Resistance of pear psylla (*Cacopsylla pyri* L.; Hom., Psyllidae) to deltamethrin and synergism with piperonyl butoxide. *J. Appl. Entomol.* **2003**, *127*, 305–312. [CrossRef]

48. Civolani, S.; Cassanelli, S.; Rivi, M.; Manicardi, G.C.; Peretto, R.; Chicca, M.; Pasqualini, E.; Leis, M. Survey of Susceptibility to Abamectin of Pear Psylla (Hemiptera: Psyllidae) in Northern Italy. *J. Econ. Entomol.* **2010**, *103*, 816–822. [CrossRef] [PubMed]

49. Bogya, S.; Marko, V.; Szinetár, C. Comparison of pome fruit orchard inhabiting spider assemblages at different geographical scales. *Agric. For. Entomol.* **1999**, *1*, 261–269. [CrossRef]

50. Solomon, M.G.; Cross, J.V.; Fitzgerald, J.D.; Campbell, C.A.M.; Jolly, R.L.; Olszak, R.W.; Niemczyk, E.; Vogt, H. Biocontrol of pests of apples and pears in northern and central Europe—3. Predators. *Biocontrol Sci. Technol.* **2000**, *10*, 91–128. [CrossRef]

51. Sanchez, J.A.; Ortín-Angulo, M.C. Abundance and population dynamics of *Cacopsylla pyri* (Hemiptera: Psyllidae) and its potential natural enemies in pear orchards in southern Spain. *Crop Prot.* **2012**, *32*, 24–29. [CrossRef]

52. Sanchez, J.A.; López-Gallego, E.; La-Spina, M. The impact of ant mutualistic and antagonistic interactions on the population dynamics of sap-sucking hemipterans in pear orchards. *Pest Manag. Sci.* **2019**, *76*, 1422–1434. [CrossRef]

53. Sanchez, J.A.; Carrasco-Ortiz, A.; López-Gallego, E.; La-Spina, M. Ants (Hymenoptera: Formicidae) reduce the density of *Cacopsylla pyri* (Linnaeus, 1761) in Mediterranean pear orchards. *Myrmecol. News* **2020**, *30*, 93–102.

54. Pérez-Marcos, M.; López-Gallego, E.; Ramírez-Soria, M.J.; Sanchez, J. Key parameters for the management and design of field margins aiming to the conservation of beneficial insects. *Landsc. Manag. Funct. Biodivers. IOBC-WPRS Bull.* **2017**, *122*, 151–155.

55. Sanchez, J.A.; Carrasco, A.; La Spina, M.; Pérez-Marcos, M.; Ortiz-Sánchez, F.J. How bees respond differently to field margins of shrubby and herbaceous plants in intensive agricultural crops of the Mediterranean area. *Insects* **2020**, *11*, 26. [CrossRef]

56. Derraik, J.G.B.; Early, J.W.; Closs, G.P.; Dickinson, K.J.M. Morphospecies and taxonomic species comparison for Hymenoptera. *J. Insect Sci.* **2010**, *10*, 1–7. [CrossRef]

57. Martínez, M.D.; Acosta, F.J.; Ruiz, E. *Claves Pra la Identificación de la Fauna Española. Las Subfamilias y Géneros de las Hormigas Ibéricas*; Universidad Complutense: Madrid, Spain, 1985.

58. Goulet, H.; Huber, J.T. *Hymenoptera of the World: An Identification Guide to Families*; Goulet, H., Huber, J.T., Eds.; Canada Communication Group: Ottawa, ON, Canada, 1993; ISBN 0660149338.

59. Nentwig, W.; Blick, T.; Bosmans, R.; Gloor, D.; Hänggi, A.; Kropf, C. Araneae Version 09.2019. Available online: https://www.araneae.nmbe.ch (accessed on 9 September 2019).

60. Salgado, J.; Outerello, R.; Gamarra, P.; Blas, M.; Vazquez, X.; Otero, J.C. Coleópteros. In *Curso Práctico de Entomología*; Barrientos, J.A., Ed.; Universitat Autònoma de Barcelona, Servei de Publicacions: Barcelona, Spain, 2004; pp. 741–811. ISBN 84-490-2383-1.

61. Bates, D.; Mächler, M.; Bolker, B.M.; Walker, S.C. Fitting linear mixed-effects models using lme4. *J. Stat. Softw.* **2015**, *67*, 1–48. [CrossRef]

62. R-Development-Core-Team. *A Language and Environment for Statistical Computing*; R Foundation for Statistical Computing: Vienna, Austria, 2017.

63. Cárdenas, M.; Castro, J.; Campos, M. Short-Term Response of Soil Spiders to Cover-Crop Removal in an Organic Olive Orchard in a Mediterranean Setting. *J. Insect Sci.* **2012**, *12*, 1–18. [CrossRef] [PubMed]

64. Rieux, R.; Simon, S.; Defrance, H. Role of hedgerows and ground cover management on arthropod populations in pear orchards. *Agric. Ecosyst. Environ.* **1999**, *73*, 119–127. [CrossRef]

65. Woolwine, A.E.; Reagan, T.E. Potential of Winter Cover Crops to Increase Abundance of *Solenopsis invicta* (Hymenoptera: Formicidae) and Other Arthropods in Sugarcane. *Environ. Entomol.* **2001**, *30*, 1017–1020. [CrossRef]

66. Tillman, G.; Schomberg, H.; Phatak, S.; Mullinix, B.; Lachnicht, S.; Timper, P.; Olson, D. Influence of cover crops on insect pests and predators in conservation tillage cotton. *J. Econ. Entomol.* **2004**, *97*, 1217–1232. [CrossRef] [PubMed]

67. Serra, G.; Lentini, A.; Verdinelli, M.; Delrio, G. Effects of cover crop management on grape pests in a Mediterranean environment. *IOBC/WPRS Bull.* **2006**, *29*, 209–214.

68. Sáenz-Romo, M.G.; Veas-Bernal, A.; Martínez-García, H.; Campos-Herrera, R.; Ibáñez-Pascual, S.; Martínez-Villar, E.; Pérez-Moreno, I.; Marco-Mancebón, V.S. Ground cover management in a Mediterranean vineyard: Impact on insect abundance and diversity. *Agric. Ecosyst. Environ.* **2019**, *283*, 106571. [CrossRef]

69. Peck, S.L.; Carolina, N. Using Ant Species as a biological indicator of Agroecosystem Condition. *Environ. Entomol.* **1998**, *27*, 1102–1110. [CrossRef]

70. Buchholz, J.; Querner, P.; Paredes, D.; Bauer, T.; Strauss, P.; Guernion, M.; Scimia, J.; Cluzeau, D.; Burel, F.; Kratschmer, S.; et al. Soil biota in vineyards are more influenced by plants and soil quality than by tillage intensity or the surrounding landscape. *Sci. Rep.* **2017**, *7*, 17445. [CrossRef]

71. Carmona, D.M.; Landis, D.A. Influence of refuge habitats and cover crops on seasonal activity-density of ground beetles (Coleoptera: Carabidae) in field crops. *Environ. Entomol.* **1999**, *28*, 1145–1153. [CrossRef]

72. Sáenz-Romo, M.G.; Veas-Bernal, A.; Martínez-García, H.; Ibáñez-Pascual, S.; Martínez-Villar, E.; Campos-Herrera, R.; Marco-Mancebón, V.S.; Pérez-Moreno, I. Effects of ground cover management on insect predators and pests in a mediterranean vineyard. *Insects* **2019**, *10*, 421. [CrossRef] [PubMed]

73. Altieri, M.A.; Schmidt, L.L. Cover Crop Manipulation in Northern California Orchards and Vineyards: Effects on Arthropod Communities. *Biol. Agric. Hortic.* **1985**, *3*, 1–24. [CrossRef]

74. Markó, V.; Keresztes, B. Flowers for better pest control? Ground cover plants enhance apple orchard spiders (Araneae), but not necessarily their impact on pests. *Biocontrol Sci. Technol.* **2014**, *24*, 574–596. [CrossRef]

75. Burgio, G.; Marchesini, E.; Reggiani, N.; Montepaone, G.; Schiatti, P.; Sommaggio, D. Habitat management of organic vineyard in Northern Italy: The role of cover plants management on arthropod functional biodiversity. *Bull. Entomol. Res.* **2016**, *106*, 759–768. [CrossRef]

76. Danne, A.; Thomson, L.J.; Sharley, D.J.; Penfold, C.M.; Hoffmann, A.A. Effects of Native Grass Cover Crops on Beneficial and Pest Invertebrates in Australian Vineyards. *Environ. Entomol.* **2010**, *39*, 970–978. [CrossRef]

77. Rodríguez, E.; González, B.; Campos, M. Natural enemies associated with cereal cover crops in olive groves. *Bull. Insectol.* **2012**, *65*, 43–49.

78. Sunderland, K.D.; Samu, F. Effects of agricultural diversification on the abundance, distribution, and pest control potential of spiders: A review. *Entomol. Exp. Appl.* **2000**, *95*, 1–13. [CrossRef]

79. Schmidt, M.H.; Roschewitz, I.; Thies, C.; Tscharntke, T. Differential effects of landscape and management on diversity and density of ground-dwelling farmland spiders. *J. Appl. Ecol.* **2005**, *42*, 281–287. [CrossRef]

80. Bauer, T. Beetles which use a setal trap to hunt springtails: The hunting strategy and apparatus of *Leistus* (Coleoptera, Carabidae). *Pedobiologia (Jena)* **1985**, *28*, 275–287.

81. Agustí, N.; Shayler, S.P.; Harwood, J.D.; Vaughan, I.P.; Sunderland, K.D.; Symondson, W.O.C. Collembola as alternative prey sustaining spiders in arable ecosystems: Prey detection within predators using molecular markers. *Mol. Ecol.* **2003**, *12*, 3467–3475. [CrossRef]

82. McIver, J.D.; Stonedahl, G. Myrmecomorphy: Morphological and behavioral mimicry of ants. *Annu. Rev. Entomol.* **1993**, *38*, 351–379. [CrossRef]

83. Pekár, S. Predatory characteristics of ant-eating *Zodarion* spiders (Araneae: Zodariidae): Potential biological control agents. *Biol. Control* **2005**, *34*, 196–203. [CrossRef]

84. Bugg, R.L.; Waddington, C. Using cover crops to manage arthropod pests of orchards: A review. *Agric. Ecosyst. Environ.* **1994**, *50*, 11–28. [CrossRef]

85. Fox, A.F.; Kim, T.N.; Bahlai, C.A.; Woltz, J.M.; Gratton, C.; Landis, D.A. Cover crops have neutral effects on predator communities and biological control services in annual cellulosic bioenergy cropping systems. *Agric. Ecosyst. Environ.* **2016**, *232*, 101–109. [CrossRef]

86. Nyffeler, M. Prey selection of spiders in the field. *J. Acharol.* **1999**, *27*, 317–324.

87. Axelsen, J.A.; Kristensen, K.T. Collembola and mites in plots fertilised with different types of green manure. *Pedobiologia (Jena)* **2000**, *44*, 556–566. [CrossRef]

88. Cobb, L.M.; Cobb, V.A. Occurrence of Parasitoid wasps, *Baeus* sp. and *Gelis* sp., in the egg sacs of the wolf spiders *Pardosa moesta* and *Pardosa sternalis* (Araneae, Lycosidae) in Southeastern Idaho. *Can. Field-Nat.* **2004**, *118*, 122–123. [CrossRef]

89. Honek, A. The Effect of Plant Cover and Weather on the Activity Density of Ground Surface Arthropods in a Fallow Field. *Entomol. Res. Org. Agric.* **1997**, *15*, 203–210. [CrossRef]

90. Diehl, E.; Wolters, V.; Birkhofer, K. Arable weeds in organically managed wheat fields foster carabid beetles by resource- and structure-mediated effects. *Arthropod Plant Interact.* **2012**, *6*, 75–82. [CrossRef]

91. Niemela, J.; Spence, J.R.; Spence, D.H. Habitat associations and seasonal activity of ground-beetles (Coleoptera, Carabidae) in Central Alberta. *Can. Entomol.* **1992**, *124*, 521–540. [CrossRef]

92. Stamps, W.T.; Nelson, E.A.; Linit, M.J. Survey of Diversity and Abundance of Ground-Dwelling Arthropods in a Black Walnut-Forage Alley-Cropped System in the Mid-Western United States. *J. Kans. Entomol. Soc.* **2009**, *82*, 46–62. [CrossRef]

93. Schipanski, M.E.; Barbercheck, M.; Douglas, M.R.; Finney, D.M.; Haider, K.; Kaye, J.P.; Kemanian, A.R.; Mortensen, D.A.; Ryan, M.R.; Tooker, J.; et al. A framework for evaluating ecosystem services provided by cover crops in agroecosystems. *Agric. Syst.* **2014**, *125*, 12–22. [CrossRef]

94. Finney, D.M.; Kaye, J.P. Functional diversity in cover crop polycultures increases multifunctionality of an agricultural system. *J. Appl. Ecol.* **2017**, *54*, 509–517. [CrossRef]

95. Corbett, A. The importance of movement in the response of natural enemies to habitat manipulation. In *Enhancing Biological Control: Habitat Management to Promote Natural Enemies of Agricultural Pests*; Pickett, C.H., Bugg, R.L., Eds.; University of California Press: Berkeley, CA, USA, 1998; pp. 25–48.

96. Wright, M.G. Cover Crops and Conservation Biocontrol: Can the Impacts of *Trichogramma* (Hymenoptera: Trichogrammatidae) Be Magnified? *Ann. Entomol. Soc. Am.* **2019**, *112*, 295–297. [CrossRef]

97. Horton, D.R.; Jones, V.P.; Unruh, T.R. Use of a new immunomarking method to assess movement by generalist predators between a cover crop and tree canopy in a pear orchard. *Am. Entomol.* **2009**, *55*, 49–56. [CrossRef]

98. Davis, A.S.; Hill, J.D.; Chase, C.A.; Johanns, A.M.; Liebman, M. Increasing Cropping System Diversity Balances Productivity, Profitability and Environmental Health. *PLoS ONE* **2012**, *7*, e47149. [CrossRef]

99. Haley, S.; Hogue, E. Ground cover influence on apple aphid, *Aphis pomi* DeGeer (Homoptera: Aphididae), and its predators in a young apple orchard. *Crop Prot.* **1990**, *9*, 225–230. [CrossRef]

100. Stephens, M.J.; France, C.M.; Wratten, S.D.; Frampton, C. Enhancing biological control of leafrollers (Lepidoptera: Tortricidae) by sowing buckwheat (*Fagopyrum esculentum*) in an orchard. *Biocontrol Sci. Technol.* **1998**, *8*, 547–558. [CrossRef]

101. Fye, R.E. Cover Crop Manipulation for Building Pear Psylla (Homoptera: Psyllidae) Predator Populations in Pear Orchards. *J. Econ. Entomol.* **1983**, *76*, 306–310. [CrossRef]

102. Pekár, S. Predatory Behavior of Two European Ant-Eating Spiders (Araneae, Zodariidae). *J. Arachnol.* **2004**, *32*, 31–41. [CrossRef]

 *agronomy*

Article

Management Intensification of Hay Meadows and Fruit Orchards Alters Soil Macro- Invertebrate Communities Differently

Elia Guariento [1,2,*], Filippo Colla [1,2], Michael Steinwandter [1], Julia Plunger [1], Ulrike Tappeiner [1,2] and Julia Seeber [1,2]

[1]   Institute for Alpine Environment, Eurac Research, Viale Druso 1, 39100 Bozen/Bolzano, Italy; filippo.colla@eurac.edu (F.C.); michael.steinwandter@eurac.edu (M.S.); julia.plunger@eurac.edu (J.P.); ulrike.tappeiner@eurac.edu (U.T.); julia.seeber@eurac.edu (J.S.)

[2]   Department of Ecology, University of Innsbruck, Technikerstrasse 25/Sternwartestrasse 15, 6020 Innsbruck, Austria

*   Correspondence: elia.guariento@eurac.edu; Tel.: +49-0471-055-298

Received: 17 April 2020; Accepted: 22 May 2020; Published: 28 May 2020

Abstract: Land-use changes and especially management intensification currently pose a major threat to biodiversity both on and beneath the soil surface. With a comparative approach, we investigated how management intensity in orchards and meadows influences soil macro-invertebrate communities in a North-Italian Alpine region. We compared soil fauna assemblies from traditional low-input sites with respective intensively managed ones. As expected, the taxonomical richness and diversity were lower in both intensive management types. Extensive management of both types revealed similar communities, while intensification led to substantial differences between management types. From these results, we conclude that intensification of agricultural practices severely alters the soil fauna community and biodiversity in general, however, the direction of these changes is governed by the management type. In our view, extensive management, traditional for mountain areas, favors soil fauna communities that have adapted over a long time and can thus be viewed as a sustainable reference condition for new production systems that consider the protection of soil diversity in order to conserve essential ecosystem functions.

Keywords: traditional management; soil biodiversity; sustainable agriculture; management intensity; South Tyrol; mountain agriculture

---

1. Introduction

The industrialization of agriculture characterized by high input practices in most cases increases production but severely hampers the overall diversity of life that inhabits managed areas from single fields to whole landscapes [1,2]. This change towards more intensive management is currently causing major losses of biodiversity on a global scale [3,4]. On the other hand, there is a growing acknowledgment that biodiversity sustains the agricultural production with several direct and indirect ecosystem services, from supporting pollination and pest control [5] to the provision of long-term stability and the maintenance of ecosystem functions [6,7]. Consequently, biodiversity loss and the resulting possible absence of ecosystem services are considered to be so profound that they pose a threat to the current and future food provisioning system [8]. Thus, developing sustainable agricultural practices that conserve biodiversity and secure the provision of food in the long run, has become a global goal and challenge [8–11]. It is of major interest (1) to describe the state of production systems that might sustain a major proportion of biodiversity and are considered sustainable long-term and (2) to investigate the changes caused by management intensification to subsequently identify and understand the potential causes of these changes.

Soils are of vital importance for agriculture by providing for the essential ecosystem functions and services, such as carbon sequestration, water regulation, soil structure alteration, nutrient cycling, and retention [12,13]. The current global loss of biodiversity is apparently happening also for soil organisms in concomitance to the general soil degradation, posing a threat to important services which soils and its inhabitants provide [14,15]. Changing land-use practices and intensifying the management contribute to the current reduction in soil biodiversity [16,17]. Recent studies on soil invertebrates found that the soil community reacts to management intensification by reducing average body size, by shortening its trophic chain length and by reducing the overall diversity, thus influencing ecosystem processes and soil quality [17–22]. However, agricultural production depends directly on the soil quality [23] and this quality is directly linked to soil inhabitants [24]. Especially macro-invertebrates are well-suited bioindicators, linked to most soil ecosystem functions and services [25], sensible to both anthropogenic and environmental changes [26] and commonly used to evaluate the general soil quality [21,27].

The present study aims to investigate the soil communities of two production systems, hay meadows and fruit orchards, both characterized by traditional, extensive, and sustainable production systems and to evaluate changes in community structures after the intensification of these management practices.

More specifically, we test (1) if taxa richness and diversity decline from extensive to intensive management; (2) if meadows and orchards harbor a different community of soil macro-invertebrates and (3) if intensification similarly influences the community of both management types.

We expect differences in the soil community between the two habitat types (meadows and orchards) since the presence of trees is known to significantly alter both biotic and abiotic characteristics in their surroundings [28]. Further, we expect that the intensification of both practices results in similar constraints for the soil community, leading to similar and less diverse community compositions [17].

2. Materials and Methods

The study region (South Tyrol, Italy) is an Alpine region where agriculture is characterized by small farms, with orchards dominating the agricultural practice in the valley floors and hay meadows on intermediate elevations [29]. Hay meadows we defined as managed grasslands mowed for hay production at least once a year, with no or limited pasturing occurring in late summer and autumn after the hay harvest.

The extensive and traditional management form of both practices was locally performed for several hundred and up to thousands of years and is expected to harbor a specialized and diverse community [30,31]. Both production types underwent land-use changes (intensification or abandonment) over the course of the last decades and intensive management has broadly replaced the extensive form, especially on lower elevations [32].

A full factorial design was chosen with a set of four fields in six different locations (Figure 1). Each location comprised two meadows and two orchards, each with intensive and extensive management; overall, 24 fields were investigated. The selection of the locations was limited by the requirement that all four management types had to be close together (based on geographical information systems). The single fields were selected on-site based both on the farmer's statement regarding the management intensity and a brief inspection of the vegetation cover. Data concerning the mowing frequencies, manuring type and frequency, grazing type, and intensity were compiled as precisely as possible (Table S1). With these parameters, a simplified version of the land-use index (LUI) following the procedure proposed by Blüthgen [33] and Fischer [34] was performed. The only difference from this approach concerned the use of manuring frequency rather than the amount of nitrogen used to fertilize since precise data could not be obtained for each field. Further, site parameters were recorded locally (coordinates, elevation, slope, exposition), while chemical parameters of soil samples (such as SOM, pH, C/N, N, and P content) were analyzed by a specialized laboratory at the Laimburg Research Center (South Tyrol). Physical parameters (soil texture) were analyzed using the automated PARIO system (meter group [35]) (for soil parameters and site characteristics see Table S1; for correlation among these factors see Figure S1). The soil samples for the chemical and physical analyses were collected in each field by taking six subsamples in 5–10 cm depth (the medium depth of the macrofauna samples) at least 10 m apart from each other (three in autumn 2018 and three in spring 2019, at the same time of sampling of the soil fauna community). These six subsamples were mixed to receive one homogenized sample per field for further analyses.

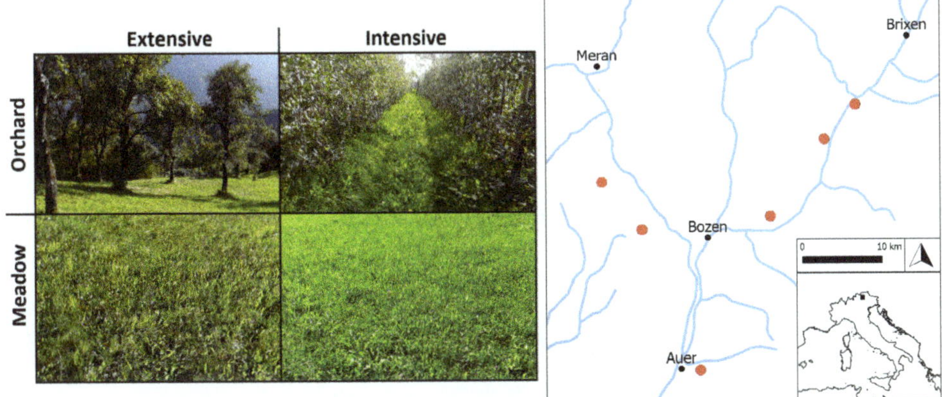

Figure 1. Schematic representation of the four differently managed field types located in each of the six sampling locations depicted in the map (red dots). The map shows a central excerpt from South Tyrol, the northernmost Province of Italy.

In total, 213 single soil core samples (20 × 20 × 15 cm square frame) were collected at three different time points (autumn 2018, spring 2019, and autumn 2019) with three sub-replicates per field and date. The three sub-replicate samples were collected within a minimum of 10 m distance from each other and in orchards within a distance of one to two meters from the tree base (e.g., for intensive apple orchards in the middle of the driveway). We added an additional sampling in autumn 2019 (except for one single field where the access was not granted by the farmer), since the very dry summer 2018 may have influenced the community composition in autumn 2018. The soil macro-invertebrates were extracted using a heat extractor (modified after [36]), quantified and identified to family level (except for Symphyla) using suitable identification keys [37–41] and following the taxonomic information of Fauna Europea Database [42]. For the ordination analysis, we treated adult and larval stages of beetles as different taxa since they differ functionally.

Sampling coverage was investigated for each single-season and field type using species accumulation curves based on the single soil core sample as sampling unites (iNext [43]; Figure S2). The three sub-replicates were then averaged to one single sample per sampling time and used as such for the following analysis (see the full community in Table S2). Taxon diversity is represented by the Shannon-Wiener diversity index (computed with function *diversity* in vegan; [44]). For the statistical analysis of taxon richness and diversity, a linear mixed model with season and regions (nested in season) modeled as random factors was constructed. Scores were extracted with the function *predict* and used to generate boxplots. Further, a mixed-model constrained ordinations (function *capscale* in vegan; [44] using Bray-Curtis distances) were used to investigate the community composition changes and a PERMANOVA (function *adonis* in vegan; [44]) was used to verify if the field type centroids differ significantly. For these computations, the community matrix was reduced by excluding ants, for which the sampling methodology is not appropriate, they differ inlife history and occurrence. All variables were standardized. The season was modeled as a random factor to account for temporal autocorrelation and the region was modeled as a random factor nested within the season to account for spatial autocorrelation between the fields. We chose single factors to be tested in the constrained ordination, which (1) represented the site characteristics, the management intensity, the chemical and the physical soil properties, (2) did not correlate significantly with each other, and (3) had *a priori* expected effect on the soil community. We tested these factors with a permutation test upon the full model and only significant parameters were implemented in the final ordination. All computations and graphics were performed and generated using the statistical programming software R [45]. Graphics were produced using the package GGPLOT2 [46], the correlation plot was generated using the package CORRPLOT [47].

3. Results and Discussion

Taxa richness (on average 18.5 ± 4.9 SD in extensive meadows; 17.1 ± 3.9 SD in intensive meadows; 19.1 ± 5.0 SD in extensive orchards; 14.1 ± 3.7 SD in intensive orchards) as well as diversity (on average 2.33 ± 0.27 SD in extensive meadows; 2.18 ± 0.17 SD in intensive meadows; 2.19 ± 0.27 SD in extensive orchards; 2.04 ± 0.25 SD in intensive orchards) was found to be significantly reduced under more intensive management in both meadows and orchards (Figure 2; Table 1). This result was expected and has already been shown for both soil [16,17,20] and above ground organisms (e.g., [4,30,48]). Nevertheless, this confirms the severe impact management intensification has on overall soil biodiversity also for the systems investigated in this study. Specifically, for meadows, there is already broad evidence that management intensification alters the soil community and its related functional aspect (e.g., [22,48]). For orchards, on the other hand, the literature is more sparse and inconsistent regarding the supported biodiversity and intensification effects, especially for soil fauna communities (but see [49,50]). Species accumulation curves depict an overall good coverage for each season and further confirm the low scoring of intensive orchards in comparison to extensive managed fields. Intensive meadows score on an intermediate position, due to a higher taxa richness in spring (Figure S2).

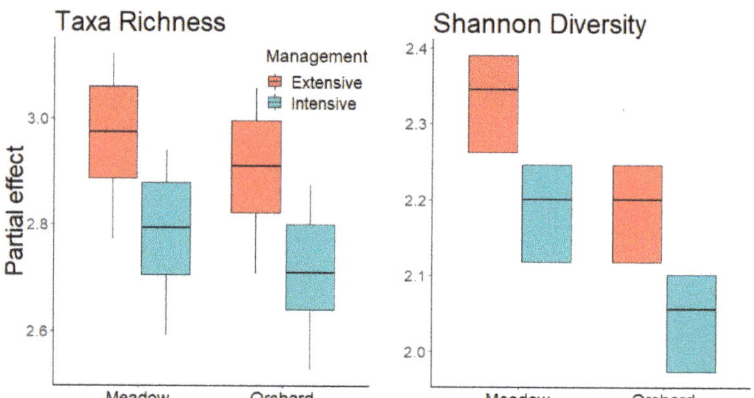

Figure 2. Boxplot of the partial effect of habitat (meadow and orchard) and management intensity (intensive and extensive) on taxa richness and diversity (Shannon diversity) (*n* = 18). The data was extracted from the linear mixed model accounting for the effect of region and season treated as random factors. Boxplot depicting the interquartile range (IQR between the 25% and the 75%) and the median, whiskers extending 1.5 * the IQR.

Table 1. Likelihood ratio test of linear mixed-model for taxa richness and diversity (Shannon diversity). Habitat (meadow vs. orchard) and management intensity (extensive vs. intensive) were tested separately. The region was implemented as a random factor nested in the season to account for both spatial and temporal autocorrelation.

Dimension	Factor	df.	Chi^2	*p*
Richness	Management	1	9.85	0.002
	Habitat	1	1.52	0.283
Diversity	Management	1	6.03	0.014
	Habitat	1	5.98	0.014

The community composition was overall significantly different between the four field types (PERMANOVA; 999 permutations: $F_{3,70}$ = 2.169, $p < 0.001$). Further, a surprisingly similar community was found in the extensively managed fields (Figure 3). This was in part unexpected because of the significant role trees are known to exert on the soil community [28]. This similarity is probably due to both the similar land-use intensity (characterized by the factor land use index (LUI) after [33]) and the generally richer and more diverse soil community adapted to the traditional extensive practice. The rather clear separate plotting of the intensive fields both from the extensive ones and from each other further underlines the important effect of management intensification and type.

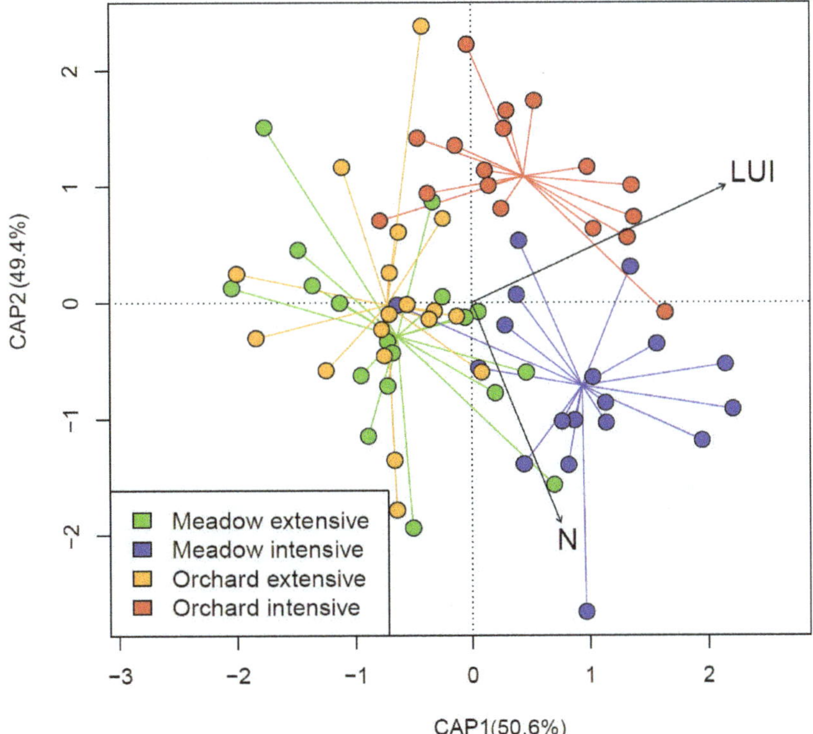

Figure 3. Ordination of the constrained canonical correspondence analysis. Each spot represents one sampling event per site. Colors and spider webs according to the field management type. Spider web centers are the weighted centroids of each management type.

Regarding site and soil parameters, we found expectable trends and correlations between the single factors (Figure S1). Only non-correlating factors representing site characteristics, physical and chemical parameters as well as management details were selected to be tested in a constrained ordination. The resulting significant parameters in a permutation test where land-use index (LUI; Permutation test: $F_{1,57} = 2.93$; $p = 0.001$) and soil nitrogen content (N; Permutation test: $F_{1,57} = 2.89$; $p = 0.001$) and were implemented in the final ordination (Biplot scores for CAP1: LUI = 0.90; N = 0.31; and for CAP2 LUI = 0.42; N = −0.78; Figure 3). The first constrained axis depicts well the management intensity, separating both the intensely managed fields from the extensive ones. The second axis clearly separates the two intensive management types from each other, contrary to our expectations (Figure 3). We expected intensification to lead to a more simple and similar community as found in several other studies [17,20,49], but this was not the case between orchards and meadows. The differences are apparently directly connected to different management practices. While the recorded management parameters and the sum factor LUI (Table S1) was not able to fully explain these differences, we suspect that (1) the more frequent presence of heavy machinery in intensive orchards (for frequent mowing, working in the fields, picking the fruits) as well as (2) the application of additional chemicals (herbicides and insecticides routinely used in the region: [51]) might further significantly impact the soil community of intensive orchards. The more frequent presence of heavy machines is probably causing soil compaction, limiting the presence of soil macro-invertebrates, and the use of additional chemicals is also known to hamper the soil biodiversity in general [15]. The quantification of both these influences has not been considered in detail in this study, but potentially explains the severe differences between intensive meadows and orchards. Further, N is highly correlating with the SOM content ($r = 0.94$) and appears to be one factor that differentiates the two intensive forms of management from each other. A decrease in SOM is generally known to be a consequence of soil degradation and a potential threat for soil biodiversity [52], accordingly, richness and diversity of soil invertebrates were lower in intensive orchards where SOM also resulted in lower values. This appears to be true for the intensive orchards but not for the intensive meadows characterized by higher nitrogen and SOM contents even exceeding those of the extensive managed fields, probably due to the higher fertilization with cattle manure.

Concluding, we found a more rich and diverse soil macro-invertebrate community in the extensive forms of the two investigated management types (viz. meadows and orchards). Further, we can state (1) that both extensive management types harbor a strikingly similar soil fauna community, (2) that intensification of land-use leads to a substantial change in these communities, and (3) that this change was profoundly influenced by the type of management.

Supplementary Materials: The following are available online at http://www.mdpi.com/2073-4395/10/6/767/s1, Figure S1: Correlation plot between all parameters. Significant correlations ($p < 0.05$) where marked with a red frame, Figure S2: Species accumulation plots where computed for each season and management type to visualize the sample coverage and species diversity more in detail, Table S1: Table with the site characteristics, Table S2: Table with taxon and site mean abundance and standard deviation for each single field. Data was standardized to 1 m^2 of sample and averaged over all nine single subplots sampled per field.

Author Contributions: Conceptualization, E.G. and J.S.; Methodology, E.G., J.S., J.P., F.C. and M.S.; Validation, J.S., F.C. and M.S.; Formal Analysis, E.G. and J.S.; Resources, U.T.; Writing—Original Draft Preparation, E.G.; Writing—Review & Editing, E.G., F.C., M.S., J.P., U.T. and J.S.; Visualization, E.G.; Supervision, U.T. and J.S.; Project Administration, J.S. All authors have read and agreed to the published version of the manuscript.

Funding: This research was funded by the Provinces of Trentino (IT), South Tyrol (IT) and the State of Tyrol (AT) and is part of the Euregio-EFH (Environment, Food and Health) project. The APC was funded by the Department of Innovation, Research and University of the Autonomous Province of Bozen/Bolzano.

Acknowledgments: We thank the numerous farmers and landowners for giving the permission to sample their fields and providing the essential information on management. We also want to thank Barbara Mosgöller and Tobias F. Weil for help in the field and in the lab. Further, we also want to thank Andreas Hilpold and Georg Niedrist for useful suggestions in site selection. Finally, we want to thank two anonymous reviewer that significantly improved the final manuscript. The authors thank the Department of Innovation, Research and University of the Autonomous Province of Bozen/Bolzano for covering the Open Access publication costs.

Conflicts of Interest: The authors declare no conflicts of interest.

References

1. Tilman, D.; Balzer, C.; Hill, J.; Befort, B.L. Global food demand and the sustainable intensification of agriculture. *Proc. Natl. Acad. Sci. USA* **2011**, *108*, 20260–20264. [CrossRef] [PubMed]
2. Gonthier, D.J.; Ennis, K.K.; Farinas, S.; Hsieh, H.Y.; Iverson, A.L.; Batáry, P.; Rudolphi, J.; Tscharntke, T.; Cardinale, B.J.; Perfecto, I. Biodiversity conservation in agriculture requires a multi-scale approach. *Proc. R. Soc. B Biol. Sci.* **2014**, *281*, 20141358. [CrossRef] [PubMed]
3. Sánchez-Bayo, F.; Wyckhuys, K.A.G. Worldwide decline of the entomofauna: A review of its drivers. *Biol. Conserv.* **2019**, *232*, 8–27. [CrossRef]
4. Seibold, S.; Gossner, M.M.; Simons, N.K.; Blüthgen, N.; Müller, J.; Ambarlı, D.; Ammer, C.; Bauhus, J.; Fischer, M.; Habel, J.C.; et al. Arthropod decline in grasslands and forests is associated with landscape-level drivers. *Nature* **2019**, *574*, 671–674. [CrossRef] [PubMed]
5. Dainese, M.; Martin, E.A.; Aizen, M.A.; Albrecht, M.; Bartomeus, I.; Bommarco, R.; Carvalheiro, L.G.; Chaplin-Kramer, R.; Gagic, V.; Garibaldi, L.A.; et al. A global synthesis reveals biodiversity-mediated benefits for crop production. *Sci. Adv.* **2019**, *5*, 554170. [CrossRef]
6. Hooper, D.U.; Adair, E.C.; Cardinale, B.J.; Byrnes, J.E.K.; Hungate, B.A.; Matulich, K.L.; Gonzalez, A.; Duffy, J.E.; Gamfeldt, L.; Connor, M.I. A global synthesis reveals biodiversity loss as a major driver of ecosystem change. *Nature* **2012**, *486*, 105–108. [CrossRef]
7. Tilman, D.; Reich, P.B.; Knops, J.M.H. Biodiversity and ecosystem stability in a decade-long grassland experiment. *Nature* **2006**, *441*, 629–632. [CrossRef]
8. FAO. *The State of the World's Biodiversity for Food and Agriculture*; Bélanger, J., Piling, D., Eds.; FAO Commission on Genetic Resources for Food and Agriculture Assessments: Rome, Italy, 2019; p. 529.
9. CBD Secretariat. *The Strategic Plan for Biodiversity 2011–2020 and the Aichi Biodiversity Targets*; Secretariat of the Convention on Biological Diversity: Nagoya, Japan, 2010.
10. Cunningham, S.A.; Attwood, S.J.; Bawa, K.S.; Benton, T.G.; Broadhurst, L.M.; Didham, R.K.; McIntyre, S.; Perfecto, I.; Samways, M.J.; Tscharntke, T.; et al. To close the yield-gap while saving biodiversity will require multiple locally relevant strategies. *Agric. Ecosyst. Environ.* **2013**, *173*, 20–27. [CrossRef]
11. Bender, S.F.; van der Heijden, M.G.A. Soil biota enhance agricultural sustainability by improving crop yield, nutrient uptake and reducing nitrogen leaching losses. *J. Appl. Ecol.* **2015**, *52*, 228–239. [CrossRef]
12. Orgiazzi, A.; Bardgett, R.D.; Barrios, E.; Behan-Pelletier, V.; Briones, M.J.I.; Chotte, J.-L.; De Deyn, G.B.; Eggleton, P.; Fierer, N.; Fraser, T.; et al. *Global Soil Biodiversity Atlas*; European Commission: Luxembourg, 2016; p. 176.
13. Hedlund, K.; Harris, J. Delivery of soil ecosystem services: From Gaia to genes. In *Soil Ecology and Ecosystem Services*; Wall, D.H., Bardgett, R.D., Behan-Pelletier, V., Herrick, J.E., Jones, T.H., Ritz, K., Six, J., Strong, D.R., van der Putten, W., Eds.; Oxford University Press: Oxford, UK, 2012; pp. 98–110.
14. Veresoglou, S.D.; Halley, J.M.; Rillig, M.C. Extinction risk of soil biota. *Nat. Commun.* **2015**, *6*, 8862. [CrossRef]
15. Nielsen, U.N. *Soil Fauna Assemblages*; Cambridge University Press: Cambridge, UK, 2019; p. 365.
16. Eggleton, P.; Vanbergen, A.J.; Jones, D.T.; Lambert, M.C.; Rockett, C.; Hammond, P.M.; Beccaloni, J.; Marriott, D.; Ross, E.; Giusti, A. Assemblages of soil macrofauna across a Scottish land-use intensification gradient: Influences of habitat quality, heterogeneity and area. *J. Appl. Ecol.* **2005**, *42*, 1153–1164. [CrossRef]
17. Tsiafouli, M.A.; Thébault, E.; Sgardelis, S.P.; de Ruiter, P.C.; van der Putten, W.H.; Birkhofer, K.; Hemerik, L.; de Vries, F.T.; Bardgett, R.D.; Brady, M.V.; et al. Intensive agriculture reduces soil biodiversity across Europe. *Glob. Chang. Biol.* **2015**, *21*, 973–985. [CrossRef] [PubMed]
18. Bai, Z.; Caspari, T.; Gonzalez, M.R.; Batjes, N.H.; Mäder, P.; Bünemann, E.K.; de Goede, R.; Brussaard, L.; Xu, M.; Ferreira, C.S.S.; et al. Effects of agricultural management practices on soil quality: A review of long-term experiments for Europe and China. *Agric. Ecosyst. Environ.* **2018**, *265*, 1–7. [CrossRef]
19. Culman, S.W.; Young-Mathews, A.; Hollander, A.D.; Ferris, H.; Sánchez-Moreno, S.; O'Geen, A.T.; Jackson, L.E. Biodiversity is associated with indicators of soil ecosystem functions over a landscape gradient of agricultural intensification. *Landsc. Ecol.* **2010**, *25*, 1333–1348. [CrossRef]

20. De Vries, F.T.; Thebault, E.; Liiri, M.; Birkhofer, K.; Tsiafouli, M.A.; Bjornlund, L.; Bracht Jorgensen, H.; Brady, M.V.; Christensen, S.; de Ruiter, P.C.; et al. Soil food web properties explain ecosystem services across European land use systems. *Proc. Natl. Acad. Sci. USA* **2013**, *110*, 14296–14301. [CrossRef]

21. Rüdisser, J.; Tasser, E.; Peham, T.; Meyer, E.; Tappeiner, U. The dark side of biodiversity: Spatial application of the biological soil quality indicator (BSQ). *Ecol. Indic.* **2015**, *53*, 240–246. [CrossRef]

22. Birkhofer, K.; Dietrich, C.; John, K.; Schorpp, Q.; Zaitsev, A.S.; Wolters, V. Regional conditions and land-use alter the potential contribution of soil arthropods to ecosystem services in grasslands. *Front. Ecol. Evol.* **2016**, *3*, 50. [CrossRef]

23. Lal, R. Restoring soil quality to mitigate soil degradation. *Sustainability* **2015**, *7*, 5875–5895. [CrossRef]

24. Parisi, V.; Menta, C.; Gardi, C.; Jacomini, C.; Mozzanica, E. Microarthropod communities as a tool to assess soil quality and biodiversity: A new approach in Italy. *Agric. Ecosyst. Environ.* **2005**, *105*, 323–333. [CrossRef]

25. Lavelle, P.; Decaëns, T.; Aubert, M.; Barot, S.; Blouin, M.; Bureau, F.; Margerie, P.; Mora, P.; Rossi, J.P. Soil invertebrates and ecosystem services. *Eur. J. Soil Biol.* **2006**, *42*, 3–15. [CrossRef]

26. Smith, J.; Potts, S.; Eggleton, P. The value of sown grass margins for enhancing soil macrofaunal biodiversity in arable systems. *Agric. Ecosyst. Environ.* **2008**, *127*, 119–125. [CrossRef]

27. Nuria, R.; Jérôme, M.; Léonide, C.; Christine, R.; Gérard, H.; Etienne, I.; Patrick, L. IBQS: A synthetic index of soil quality based on soil macro-invertebrate communities. *Soil Biol. Biochem.* **2011**, *43*, 2032–2045. [CrossRef]

28. Barrios, E.; Sileshi, G.W.; Shepherd, K.; Sinclair, F. Agroforestry and soil health: Linking trees, soil biota, and ecosystem services. In *Soil Ecology and Ecosystem Services*; Wall, D.H., Bardgett, R.D., Behan-Pelletier, V., Herrick, J.E., Jones, T.H., Ritz, K., Six, J., Strong, D.R., van der Putten, W., Eds.; Oxford University Press: Oxford, UK, 2012; pp. 315–330.

29. Kuttner, M.; Essl, F.; Peterseil, J.; Dullinger, S.; Rabitsch, W.; Schindler, S.; Hülber, K.; Gattringer, A.; Moser, D. A new high-resolution habitat distribution map for Austria, Liechtenstein, southern Germany, South Tyrol and Switzerland. *J. Protec. Mt. Areas Res. Manag.* **2015**, *7*, 18–29. [CrossRef]

30. Unterluggauer, P. Traditionell und intensiv bewirtschaftete wiesen in südtirol-ihre vegetation als indikator für die bewirtschaftungsintensität vegetation types as estimates of the management intensity. *Tuexenia* **2016**, *36*, 37–62.

31. Hilpold, A.; Seeber, J.; Fontana, V.; Niedrist, G.; Steinwandter, M.; Tasser, E.; Tappeiner, U. Decline of rare and specialized species across multiple taxonomic groups after grassland intensification and abandonment. *Biodivers. Conserv.* **2018**, *14*, 3729–3744. [CrossRef]

32. Egarter Vigl, L.; Schirpke, U.; Tasser, E.; Tappeiner, U. Linking long-term landscape dynamics to the multiple interactions among ecosystem services in the European Alps. *Landsc. Ecol.* **2016**, *31*, 1903–1918. [CrossRef]

33. Blüthgen, N.; Dormann, C.F.; Prati, D.; Klaus, V.H.; Kleinebecker, T.; Hölzel, N.; Alt, F.; Boch, S.; Gockel, S.; Hemp, A.; et al. A quantitative index of land-use intensity in grasslands: Integrating mowing, grazing and fertilization. *Basic Appl. Ecol.* **2012**, *13*, 207–220. [CrossRef]

34. Fischer, M.; Bossdorf, O.; Gockel, S.; Hänsel, F.; Hemp, A.; Hessenmöller, D.; Korte, G.; Nieschulze, J.; Pfeiffer, S.; Prati, D.; et al. Implementing large-scale and long-term functional biodiversity research: The biodiversity exploratories. *Basic Appl. Ecol.* **2010**, *11*, 473–485. [CrossRef]

35. Durner, W.; Iden, S.C.; von Unold, G. The integral suspension pressure method (ISP) for precise particle-size analysis by gravitational sedimentation. *Water Resour. Res.* **2017**, *53*, 33–48. [CrossRef]

36. Kempson, D.; Lloyd, M.; Ghelardi, R. A new extractor for woodland litter. *Pedobiologia* **1963**, *3*, 1–21.

37. Schaefer, M. *Brohmer-Fauna von Deutschland: Ein Bestimmungsbuch Unserer Heimischen Tierwelt*; Quelle & Meyer Verlag: Wiebelsheim, Germany, 2017; p. 765.

38. Klausnitzer, B.; Hannemann, H.-J.; Senglaub, K. *Exkursionsfauna von Deutschland: Band 2 Wirbellose: Insekten*; Springer: Dresden, Germany, 2011; p. 937.

39. Hannemann, H.-J.; Klausnitzer, B.; Senglaub, K. *Stresemann-Exkursionsfauna von Deutschland. Band 1: Wirbellose (Ohne Insekten)*; Springer: Dresden, Germany, 2019; p. 673.

40. Smith, K.G.V. An introduction to the immature stages of British flies: *Diptera larvae*, with notes on eggs, *Puparia* and *Pupae*. *Handb. Identif. Br. Insects* **1989**, *10*, 280.

41. Latella, L.; Gobbi, M. *La Fauna del Suolo: Tassonomia, Ecologia e Metodi di Studio dei Principali Gruppi di Invertebrati Terrestri Italiani*; Quaderni del Museo delle Scienze: Trento, Italy, 2015; p. 205.

42. De Jong, Y.; Verbeek, M.; Michelsen, V.; de Bjørn, P.P.; Los, W.; Steeman, F.; Bailly, N.; Basire, C.; Chylarecki, P.; Stloukal, E.; et al. Fauna Europaea-All European animal species on the web. *Biodivers. Data J.* **2014**, *2*, e4034. [CrossRef] [PubMed]

43. Hsieh, T.C.; Ma, K.H.; Chao, A. iNEXT: An R package for rarefaction and extrapolation of species diversity (hill numbers). *Methods Ecol. Evol.* **2016**, *7*, 1451–1456. [CrossRef]

44. Oksanen, J.; Blanchet, F.G.; Friendly, M.; Kindt, R.; Legendre, P.; McGlinn, D.; Minchin, P.R.; O'Hara, R.B.; Simpson, G.L.; Solymos, P.; et al. Vegan: Community Ecology Package. R Package Version 2.5–2. Available online: https://CRAN.R-project.org/package=vegan (accessed on 17 April 2020).

45. Team, R.C. *R: A Language and Environment for Statistical Computing*; R Foundation for Statistical Computing: Vienna, Austria, 2017.

46. Wickham, H. *ggplot2: Elegant Graphics for Data Analysis*, 2nd ed.; Springer: New York, NY, USA, 2016; p. 260.

47. Wei, T.; Simko, V.; Levy, M.; Xie, Y.; Jin, Y.; Zemla, J. Package "corrplot". *Statistician* **2017**, *56*, 316–324.

48. Simons, N.K.; Weisser, W.W.; Gossner, M.M. Multi-taxa approach shows consistent shifts in arthropod functional traits along grassland land-use intensity gradient. *Ecology* **2016**, *97*, 754–764. [CrossRef] [PubMed]

49. Mazzia, C.; Pasquet, A.; Caro, G.; Thénard, J.; Cornic, J.F.; Hedde, M.; Capowiez, Y. The impact of management strategies in apple orchards on the structural and functional diversity of epigeal spiders. *Ecotoxicology* **2015**, *24*, 616–625. [CrossRef]

50. Paoletti, M.G.; Schweigl, U.; Favretto, M.R. Soil macroinvertebrates, heavy metals and organochlorines in low and high input apple orchards and a coppiced woodland. *Pedobiologia* **1995**, *39*, 20–33.

51. Dalla Via, J.; Mantinger, H. Die Landwirtschaftliche forschung im obstbau südtirols: Entwicklung und ausblick. *Erwerbs-Obstbau* **2012**, *54*, 83–115. [CrossRef]

52. Gardi, C.; Jeffery, S.; Saltelli, A. An estimate of potential threats levels to soil biodiversity in EU. *Glob. Chang. Biol.* **2013**, *19*, 1538–1548. [CrossRef]

MDPI

St. Alban-Anlage 66

4052 Basel

Switzerland

Tel. +41 61 683 77 34

Fax +41 61 302 89 18

www.mdpi.com

Agronomy Editorial Office

E-mail: agronomy@mdpi.com

www.mdpi.com/journal/agronomy